Mechanik, Werkstoffe und Konstruktion im Bauwesen

Band 53

Reihe herausgegeben von

Ulrich Knaack, Darmstadt, Deutschland

Jens Schneider, Darmstadt, Deutschland

Johann-Dietrich Wörner, Darmstadt, Deutschland

Stefan Kolling, Gießen, Deutschland

Institutsreihe zu Fortschritten bei Mechanik, Werkstoffen, Konstruktionen, Gebäudehüllen und Tragwerken. Das Institut für Statik und Konstruktion der TU Darmstadt sowie das Institut für Mechanik und Materialforschung der TH Mittelhessen in Gießen bündeln die Forschungs- und Lehraktivitäten in den Bereichen Mechanik, Werkstoffe im Bauwesen, Statik und Dynamik, Glasbau und Fassadentechnik, um einheitliche Grundlagen für werkstoffgerechtes Entwerfen und Konstruieren zu erreichen. Die Institute sind national und international sehr gut vernetzt und kooperieren bei grundlegenden theoretischen Arbeiten und angewandten Forschungsprojekten mit Partnern aus Wissenschaft, Industrie und Verwaltung. Die Forschungsaktivitäten finden sich im gesamten Ingenieurbereich wieder. Sie umfassen die Modellierung von Tragstrukturen zur Erfassung des statischen und dynamischen Verhaltens, die mechanische Modellierung und Computersimulation des Deformations-, Schädigungs- und Versagensverhaltens von Werkstoffen, Bauteilen und Tragstrukturen, die Entwicklung neuer Materialien, Produktionsverfahren und Gebäudetechnologien sowie deren Anwendung im Bauwesen unter Berücksichtigung sicherheitstheoretischer Überlegungen und der Energieeffizienz, konstruktive Aspekte des Umweltschutzes sowie numerische Simulationen von komplexen Stoßvorgängen und Kontaktproblemen in Statik und Dynamik.

Weitere Bände in der Reihe http://www.springer.com/series/13824

Felix Dillenberger

On the anisotropic plastic behaviour of short fibre reinforced thermoplastics and its description by phenomenological material modelling

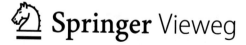 Springer Vieweg

Felix Dillenberger
Mainz, Germany

Vom Fachbereich 13 – Bau- und Umweltingenieurwissenschaften der Technischen Universität Darmstadt zur Erlangung des akademischen Grades eines Doktor-Ingenieurs (Dr.-Ing.) genehmigte Dissertation von M. Sc. Felix Ben Dillenberger aus Al Hoceima

1. Gutachten: Prof. Dr.-Ing. Jens Schneider
2. Gutachten: Prof. Dr.-Ing. habil. Stefan Kolling

Tag der Einreichung: 14.01.2019
Tag der mündlichen Prüfung: 11.04.2019
Darmstadt 2019

D17

ISSN 2512-3238   ISSN 2512-3246   (electronic)
Mechanik, Werkstoffe und Konstruktion im Bauwesen
ISBN 978-3-658-28198-4   ISBN 978-3-658-28199-1   (eBook)
https://doi.org/10.1007/978-3-658-28199-1

This Springer Vieweg imprint is published by the registered company Springer Fachmedien Wiesbaden GmbH part of Springer Nature.
The registered company address is: Abraham-Lincoln-Str. 46, 65189 Wiesbaden, Germany

Danksagung

Die vorliegende Arbeit entstand während meiner Tätigkeit am Bereich Kunststoffe des Fraunhofer-Instituts für Systemzuverlässigkeit und Betriebsfestigkeit LBF in Darmstadt in Kooperation mit dem Institut für Statik und Konstruktion (ISMD) der Technischen Universität Darmstadt und dem Institut für Mechanik und Materialforschung (IMM) der Technischen Hochschule Mittelhessen.

Herrn Prof. Dr.-Ing. habil. Stefan Kolling und Herrn Prof. Dr.-Ing. Jens Schneider gilt mein Dank für die Übernahme der Betreuung meiner Arbeit. Ihnen und Herrn Prof. Dr.-Ing. Ulrich Knaack sei gedankt für die konstruktiven Anregungen sowie die wertvollen Impulse auf den geschätzten Doktorandenseminaren. Besonders Stefan danke ich für die intensive Unterstützung und die vielen fruchtbaren Diskussionen.

Ohne meine Kollegen wäre die Arbeit in der vorliegenden Form nicht möglich gewesen. Mein Dank gilt Axel Nierbauer, Reinhold Riesner und Benedikt Leibold, die mir bei allen praktischen Fragen eine große Hilfe waren. Dr.-Ing. Sascha Sedelmeier, Dr.-Ing. Alexander Knieper und Felix Weidmann sei gedankt für die gegenseitige Motivation. Ihnen und allen weiteren Mitarbeitern des LBFs sowie des ehemaligen Deutschen Kunststoff-Instituts danke ich für viele hilfreiche Diskussionen und den kollegialen Rahmen. Ich danke zudem allen Mitarbeitern des ISMD und des IMM für den produktiven Austausch in den letzten Jahren.

Darüber hinaus gilt mein Dank allen, die mich im Rahmen eines Lektorates unterstützt haben. Ich danke hier, neben bereits genannten Personen, Stefan Williams, Benedikt Hülpüsch und Dr.-Ing. Christian Beinert. Papa und Stephan, vielen Dank für eure unermessliche Mühe.

Insbesondere meinen Eltern möchte ich danken. Sie haben mich stets uneingeschränkt unterstützt und gefördert und mir dadurch den Weg zu dieser Arbeit erst möglich gemacht. Weiterhin danke ich meinen Geschwistern, die mich in jeder Phase begleitet haben.

Meine Familie ist mir die wertvollste Stütze und hat mich im Rahmen der Promotion immer wieder gelehrt die wichtigen Dinge nicht aus den Augen zu verlieren. Danke Karl und Klara, dass ihr auf Papazeit verzichtet habt. Danke Katja für Deine unendliche Geduld.

Mainz, August 2019

Felix Ben Dillenberger

Zusammenfassung

Für die sichere Entwicklung von thermoplastischen Bauteilen muss deren mechanisches Verhalten durch Simulationen mittels der Finite-Elemente-Methode (FEM) möglichst genau vorhergesagt werden können. Ein typisches Beispiel für die Auslegung solcher Bauteile findet sich im Entwicklungsprozess für Komponenten in Automobilanwendungen in Bezug auf Crash, Insassen- und Fußgängerschutz. Hierzu sind adäquate Materialmodelle unverzichtbar. In diesem Zusammenhang wird in dieser Arbeit ein neuartiger Materialmodellierungsansatz für kurzglasfaserverstärkte Thermoplaste (KFVTP) vorgestellt.

KFVTP bestehen aus einer thermoplastischen Matrix die durch Kurzglasfasern verstärkt wird, um die Steifigkeits- und Festigkeitseigenschaften zu verbessern. KFVTP Bauteile werden kostengünstig im Spritzguß verarbeitet und weisen aufgrund der komplexen, lokal inhomogenen Faserverteilung, die während des Spritzgießens entsteht, ein unterschiedliches Maß an Anisotropie auf. Durch den Einfluss der Matrix zeigen KFVTP zudem ein nichtlineares visko-plastisches Verhalten. Die plastische Verformung und das Materialversagen sind stark abhängig von der aufgebrachten Last bzw. dem Spannungszustand im Bauteil.

Das im Rahmen dieser Arbeit entwickelte Materialmodell berücksichtigt die zuvor genannten Effekte und ist für explizite FEM entwickelt worden. Es kann linear-elastisches Verhalten und nichtlineare, orthotrope und visko-plastische Verformung von KFVTP abhängig vom Spannungszustand abbilden. Ziel ist es, einen vereinfachten Ansatz zur Vorhersage des komplexen Materialverhaltens in der technischen Praxis aufzuzeigen.

Hinsichtlich der Elastizität und der Proportionalitätsgrenze wird ein mikromechanisches Modell auf Basis der Mori-Tanaka Homogenisierung angewendet. Dabei wird auf Matrixebene eine isotrope quadratische Fließfläche eingeführt, die auf Composite-Ebene in eine Tsai-Wu Fließfläche überführt wird. Das nichtlineare Verhalten kann zu großen Teilen aus einem einzigen Experiment für jeden Spannungszustand bestimmt werden. Hierzu wird eine analytische Gleichung zur Beschreibung der nichtlinearen Verfestigung vorgestellt. Anhand von experimentellen Ergebnissen wird gezeigt, dass eine vereinfachte Abbildung der plastischen Dehnratenabhängigkeit möglich ist. Das vorgestellte Materialmodell enthält weiterhin eine skalare Formulierung für elastische Schädigung und eine auf dem Tsai-Wu Modell basierende Versagensdefinition.

Insgesamt ermöglicht das konstitutive Modell so die Berechnung des KFVTP-Verhaltens für beliebige Spannungszustände und Faserverteilungen. Dies wurde anhand von experimentellen Ergebnissen eines selbst compoundierten PPGF30 Materials hinsichtlich Zug, Druck und Schub für verschiedene Orientierungsverteilungen verifiziert.

Abstract

A requirement for the safe design of thermoplastic parts is the ability to precisely predict mechanical behaviour by finite element (FE) simulations. Typical examples for the design of such parts include the engineering of relevant components in automotive applications regarding crash, occupant- and pedestrian safety. For this purpose adequate material models are essential.

In this context, the present work introduces a novel material modelling approach for short fibre reinforced thermoplastics (SFRTPs).

SFRTPs consist of a thermoplastic matrix that is reinforced by short glass fibres in order to enhance stiffness and strength properties. SFRTP components are processed cost-effectively by injection moulding. They show a varying degree of anisotropy due to the complex locally inhomogeneous fibre distributions that arise during the injection moulding process. Moreover, the influence of the thermoplastic matrix leads to a non-linear viscoplastic behaviour. Plastic deformation and failure of SFRTPs depends strongly on the applied load, respectively the stress state that arises in the component.

The novel material model presented here takes into account all of these effects and was developed for explicit FE-Simulations. It is capable of representing linear-elastic behaviour and the non-linear orthotropic stress state dependent viscoplastic deformation of SFRTPs. It is aimed at presenting a simplified approach for the prediction of complex material behaviour in engineering practice.

Regarding elasticity and the initial yield surface, a micro-mechanical Mori-Tanaka based approach is applied. Within this approach, on the matrix level an isotropic quadratic yield surface is introduced which is transferred into a Tsai-Wu yield surface on the composite level. Furthermore, non-linear behaviour is determined for a large part from a single experiment for each stress state. In this framework, an analytical equation for the description of the non-linear hardening of SFRTPs is presented. Based on experiments it is shown that a simplified approach regarding strain rate dependency is possible. Additionally, the material model developed in this work contains a scalar damage model and a Tsai-Wu based failure definition.

Overall, the constitutive model allows for the computation of SFRTP behaviour for arbitrary stress states and orientation distributions. This was verified with the experimental results of a self-compounded PPGF30 material regarding tension, compression and shear for different orientation distributions.

Contents

Glossaries

Abbreviations

General

2D	two dimensional
3D	three dimensional
AiF	German Federation of Industrial Research Associations/ Arbeitsgemeinschaft industrieller Forschungsvereinigungen "Otto von Guericke" e.V.
CCD	charge-coupled device sensor for digital image acquisition
CPRMA	cutting plane return mapping algorithm
DIC	digital image correlation
FAD	fibre aspect ratio distribution
FEA	finite element analysis
FLD	fibre length distribution
FOD	fibre orientation distribution
ITM	Chair for Continuum Mechanics at Karlsruhe Institute of Technology
LBF	Fraunhofer Institute for Structural Durability and System Reliability LBF
ODF	orientation distribution function
SAMP	semi-analytical model for polymers
UD	unidirectional

Materials

PA	polyamide
PP	polypropylene
PPCF10	polypropylene reinforced with 10 wt.% short carbon fibres
PPCF20	polypropylene reinforced with 20 wt.% short carbon fibres
PPGF30	polypropylene reinforced with 30 wt.% short glass fibres
PTFE	polytetrafluoroethylene

| SFRTP | short fibre reinforced thermoplastic |

Testing Equipment

DMA	dynamic mechanical analysis
HTM5020	high speed servo-hydraulic HTM materials testing machine for up to 50 kN by Zwick Roell AG
μCT	micro computed tomography
SEM	scanning electron microscope
Z005	static zwickiLine materials testing machine for up to 5 kN by Zwick Roell AG
Z020	static AllroundLine universal materials testing machine for up to 20 kN by Zwick Roell AG
Z250	static AllroundLine universal materials testing machine for up to 250 kN by Zwick Roell AG

Symbols

Mechanics

α	anisotropy degree
A	cross-sectional area of the specimen
$\mathbf{A}$	orientation tensor
β	load relation factor
$\mathbf{B}$	left Cauchy-Green stretch tensor
$\mathbf{C}$	right Cauchy-Green stretch tensor
$\mathbb{C}$	stiffness tensor
d	damage variable
D	damage matrix
δ	symmetrical identity tensor
$\vec{e}$	basis vector
ε	strain
E	Young's modulus
F	Force
f	scaling factor
$\mathbf{F}$	deformation gradient, or parameters of the Tsai-Wu yield surface in stress space
G	shear modulus
h	flow potential

H	displacement gradient, or
	parameters of the flow potential in stress space
I	tensor invariant
κ	triaxiality factor, measure of multi-axiality
K	parameters of the Tsai-Wu yield surface in strain space
l	specimen length
L	velocity gradient
λ	eigenvalue
$d\lambda$	plastic multiplier
L	Eshelby tensor
$\vec{n}$	normal vector
η	viscosity
N	total number of items
p	hydrostatic pressure
$\vec{p}$	orientation vector
ϕ	planar angle
φ	aspect ratio
Φ	yield surface
P	point on a body
P	testing position
$\mathbb{P}$	stress concentration tensor
Ψ	orientation distribution function
ψ	internal hardening variable
$\vec{q}$	eigenvector
$\mathbb{Q}$	strain concentration tensor
r	radius
ρ	density
R	rotation tensor
s	specimen thickness
$\boldsymbol{\sigma}$	stress
$\mathbb{S}$	compliance tensor
ϑ	volume fraction
θ	spatial angle
T_g	glass transition temperature
$\vec{t}$	stress vector
$\mathbb{T}$	special stress concentration tensor
t	time
$\vec{u}$	displacement
U	right stretch tensor
V	volume

ν	Poisson's ratio
$\mathbf{V}$	left stretch tensor
$\mathbf{W}$	rate-of-rotation tensor
$\vec{x}$	position vector
ξ	yield surface parameter
X	failure value
Y	yield strength
ς_i	auxiliary variables
χ_i	independent variables of transversely isotropic stiffness matrix
ζ	material model parameter

Subscripts

0	initial or reference state
F	failure value
M	experimental maximum value
max	fitted asymptotic maximum value
QS	quasi-static
rp	offset value
Y	yield value

Superscripts

2D	based on two dimensional strain evaluation
b	biaxial
c	uniaxial compression
c	reinforced composite property
f	fibre property
i	property of subset i
m	polymeric matrix property
s	shear
t	uniaxial tension
dev	deviatoric quantity
el	elastic
eng	engineering quantity
isoc	isochoric
CS	based on the Cowper-Symonds model
DP	based on the Drucker-Prager model
expon	based on the exponential model
EY	based on Eshelby's model
GJ	based on the Gsell-Jonas model

H	based on the Hill yield surface model
JC	based on the Johnson-Cook model
MT	based on the Mori-Tanaka model
qdr	based on a quadratic fit
RB	based on Reuss's bound assumption
vM	based on the von Mises model
SAMP	based on the SAMP model
S	based on the Schmachtenberg model
UD	unidirectional quantity
VB	based on Voigt's bound assumption
pl	plastic
sym	symmetrical quantity
vp	viscoplastic
□	specimen from standard plate
⫾	specimen from flat bar

Notation

In the present work the following notation is used: Standard non-bold symbols refer to scalar values. Roman symbols with superscript arrows refer to first order tensors (e.g. $\vec{T}$). Simple roman bold symbols depict tensors of second order (e.g. $\mathbf{T}$). Fraktur letters illustrate tensors of third order (e.g. $\mathfrak{T}$) and tensors of fourth order are represented by the use of roman blackboard bold letters (e.g. $\mathbb{T}$). In case reduced notation is used, bold italic symbols (e.g. $\boldsymbol{T}$) represent the reduced (6×6) matrix notation of a fourth order tensor and italic symbols with superscript arrows (e.g. $\vec{T}$) depict the reduced (6×1) vector notation of a second order tensor.

The double contraction tensor operation is represented by a colon (:). The dyadic tensor product is depicted by a $\otimes$ symbol. A scalar product is labelled without an operator. In case of an inner tensor product or a matrix product, no operator symbol is applied.

1 Motivation and Outline

Plastic components increasingly substitute parts traditionally made of metallic materials in mechanical applications. This is due to an immanent lightweight potential and the possibility of tailoring physical properties to specific applications. Mechanical loading capabilities of plastic materials can be increased significantly by reinforcement fibres. The main advantages of the resulting composite materials are higher stiffness and tensile strength, low weight, favourable abrasion and wear properties as well as a low coefficient of thermal expansion. These properties can be modified to meet the requirements of specific applications by altering the formulation of the matrix materials, by adjusting the orientation of the reinforcement fibres and by taking advantage of the anisotropic nature of the composite load bearing capabilities.

Highest stiffness and strength demands can be met by continuous fibres embedded in epoxy matrices, typically applied as laminates, but production of these composites is considered time, cost and labour intensive, cf. PAPATHANASIOU et al., 1997. Moreover, these materials show a highly brittle fracture behaviour which is problematic in applications that require energy absorption capacity.

In mechanical structures which are less demanding in terms of stiffness and maximum strength, thermoplastic matrix materials with discontinuous fibre reinforcement have gained technological importance. The advantage of these materials, as opposed to continuous fibre reinforced composites, is the applicability of polymer processing techniques such as injection moulding, resulting in low costs at high throughput, an improved surface quality as well as the possibility of intricately shaped parts. Using these processing techniques, parts can be flexibly designed, since a large variety of shapes is possible. One class of discontinuous fibre reinforcements are the so-called short fibres, which consist of a variable length of up to 1 mm, cf. SCHÜRMANN, 2007. Thermoplastic composites enhanced by such fibres are utilised for a multitude of technical problems and have found wide use in automotive applications, the field of civil engineering, the aerospace sector, marine applications, as well as household products and electronics, compare FU et al., 2009.

The resulting mechanical behaviour of composites depends on the physical properties of the constituents involved. Composites show intermediate properties that range between those of the thermoplastic matrix and those of the of pure fibres. Besides being influenced by filler content and form, the distribution of fibres in the polymeric bulk material affects the overall performance. A characteristic of discontinuous reinforcement is that fibres are usually imperfectly distributed and aligned, as a result of the injection moulding

© Springer Fachmedien Wiesbaden GmbH, part of Springer Nature 2020
F. Dillenberger, *On the anisotropic plastic behaviour of short fibre reinforced thermoplastics and its description by phenomenological material modelling*, Mechanik, Werkstoffe und Konstruktion im Bauwesen 53, https://doi.org/10.1007/978-3-658-28199-1_1

production process, as shown in e.g. KLUSEMANN et al., 2009. Injection moulding process induced micro-structural property distributions may locally vary and generally leads to a complex anisotropic material behaviour, regarding, f.e., local stiffness, strength and fracture properties, see e.g. STOMMEL et al., 2018. Short fibre reinforced thermoplastic (SFRTP) composites show similarities to concrete reinforced with chopped steel fibres, as described in e.g. EIK, 2014.

In order to assure a safe operation of SFRTP components during loading, knowledge and adequate prognosis of mechanical part behaviour is crucial. While stiffness and strength properties of thermoplastics are improved by fibre reinforcement, the overall straining capability of these materials is drastically decreased. Whereas non-reinforced thermoplastics generally show ductile mechanical behaviour, composites with significant fibre fractions show brittle fracture characteristics. Therefore, an adequate prediction of their critical deformation limit is imperative when the full potential of composite materials is to be utilised in part design. Properly prognosticating material fracture requires the accurate description of the overall process preceding material deformation. This includes elastic characteristics, as well as the onset and the progress of plastic deformation. In complex loading situations where analytical evaluation of mechanical behaviour reaches its limits, prediction of mechanical part behaviour is typically done by means of numerical simulations using finite element analysis (FEA). Numerical methods in the framework of FEA have gained large importance in different fields of engineering. Therefore, reasonable simulation models of parts and material behaviour are required, cf. STOMMEL et al., 2018.

SFRTP materials call for sophisticated material models due to their anisotropic nature and the complex mechanical characteristics of the included thermoplastics. Mechanical behaviour of thermoplastics strongly differs from the performance of metals, see e.g. RÖSLER et al., 2008. Experiments to derive material data and extract material parameters have to be adjusted to the specific material models. Overall, an adequate strategy must be available for the prediction of complex mechanical SFRTP component behaviour with sufficient accuracy for the constructive design of components in a product development process.

Plenty of researchers have assessed the problem of material models for SFRTPs and the methods proposed can essentially be classified into two strategies. One of these is termed the phenomenological approach, wherein composite behaviour related to a specific local fibre orientation is mathematically described as a whole. A large part of the phenomenological approaches are derived from models used in metal applications. Thus, phenomenological models are mostly well established and easily applicable. Their computational efficiency is high. However, the necessary assessment of variable fibre orientation distributions (FODs) in SFRTP parts leads to a considerable experimentation effort. This is due to the fact that a concise FOD integration strategy is still a topic of ongoing research. The second modelling approach is the fully coupled micro-mechanical strategy, which takes into account the fibre and the matrix properties separately. While respective elastic predictions are highly accurate, the consideration of plasticity leads to complex and computationally demanding model expressions. The experimental effort on the other hand is relatively limited. A detailed account on the state of the art of SFRTP material modelling is given in chapter 2.6.

In the framework of this work, the overall material modelling process for SFRTP parts is investigated and suggestions considering the improvement of the process are made. The need was identified for a material description that combines the ease of use and versatility of micro-mechanical models with the computational advantages of phenomenological approaches. Therefore, a new material model was developed and implemented for explicit FEA. The model is heavily based on a detailed experimental evaluation of SFRTP behaviour. Experimental findings allowed to determine the relevance of including specific mechanical aspects in the material model in the context of an adequate prediction of the overall composite response. This especially involved the relationship between FOD and effective mechanical behaviour. Using the experimental results, a simplified modelling approach was derived. Overall, this work gives a guideline to a simplified simulation strategy considering the modelling of SFRTP in engineering tasks. It provides the possibility of predicting the mechanical behaviour of short glass fibre reinforced thermoplastics, with a special focus on crash applications. The derived material model offers the capability of describing thermoplastic components with varying filler content for different FODs and is applicable in preliminary as well as detailed design phases.

Most of the experimental results presented in this research have been evaluated as part of the following research projects at the Fraunhofer Institute for Structural Durability and System Reliability LBF:

- "Finite-Element-Validation - A New Method for Finite-Element-Validation in Crashworthiness Analysis using Optical Sensors", DILLENBERGER, 2014, supported under grant number 59 EN by the German Federation of Industrial Research Associations/ Arbeitsgemeinschaft industrieller Forschungsvereinigungen "Otto von Guericke" e.V. (AiF) in the context of the Industrial Collective Research (IGF) program, funded by the German Federal Ministry of Economic Affairs and Energy on the basis of a decision by the German Bundestag.

- "Modellierung der orientierungsabhängigen mechanischen Eigenschaften von kohlenstofffaserverstärkten Thermoplastformteilen", AMBERG et al., 2013, supported under grant number 16712N by AiF.

- "Phänomenologische Berechnungsstrategie für kurzfaserverstärkte Spritzgussform-teile", FORNOFF et al., 2018, supported under grant number 18362N by AiF.

The results presented in this research therefore have partially been previously publicised in the corresponding reports DILLENBERGER, 2014, AMBERG et al., 2013 and FORNOFF et al., 2018. Additionally, certain results have been published in the research paper MÜLLER et al., 2015.

Figure 1.1 shows the structure of the present work. Chapter 2 presents the state of the art considering mechanical behaviour of SFRTP materials. In the first sections of this chapter an overview of mechanical properties, definitions and general behaviour of thermoplastic SFRTP materials is given. It includes a state of the art review regarding the description and determination of fibre distributions. The chapter is concluded by a broad state of the art review of current research on SFRTP material modelling in order to classify this work and substantiate the remarks given in the previously stated motivation. Chapter 3 introduces basic mechanical definitions used throughout this work. It is followed by chapter 4 that focuses on the derivation and evaluation of experimental data. A detailed material characterisation was conducted with respect to possible applications in FEA. The testing program comprised mechanical behaviour of polypropylene reinforced with 30 wt.% short glass fibres (PPGF30) under different loading velocities, loading directions and stress states. Fibre configurations induced by injection moulding were evaluated by micro-structural analysis, in order to determine the relationship between fibre properties and effective material behaviour. Subsequently, the modelling of the relevant aspects of the mechanical composite behaviour found in the experiments is discussed in chapter 5. It presents the constitutive equations and an algorithm for a material model applicable in

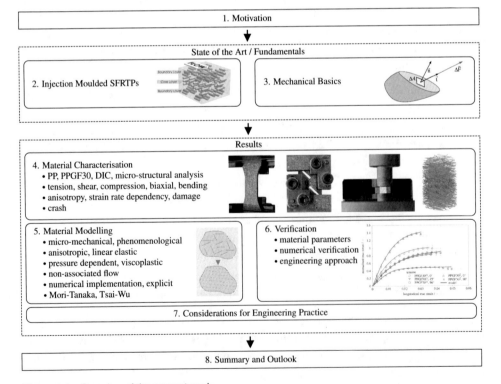

Figure 1.1   Overview of the present work.

explicit FE-Simulations. In chapter 6 previous assumptions and modelling expressions are verified against experimental data. Chapter 7 represents a synopsis of recommendations for the application of the results in current engineering practice and outlines suggestions on simplifications that can be made in order to reduce the complexity of the material model parametrisation. Finally chapter 8 gives a summary of the findings of the present work and a conclusion. It includes an outlook on the possible further extension of the results of this work.

2 Injection Moulded Short Fibre Reinforced Thermoplastics

This work considers materials that can be distinguished as short fibre reinforced thermo-plastics (SFRTPs), processed by injection moulding. They consist of a thermoplastic matrix material reinforced by chopped short glass fibres. The following sections give a general overview of these types of materials. More detailed accounts on the characteristics of polymer materials can be found in textbooks, e.g. DOMININGHAUS, 2008, BONNET, 2009, EYERER et al., 2008 among others, as well as references given in the respective sections.

Figure 2.1 Exemplary micrograph of a fracture surface of a short fibre reinforced thermoplastic generated by scanning electron microscope (SEM).

Figure 2.1 shows an exemplary fracture surface of an SFRTP. The large plastic deforma-tion of the matrix material into which the fibres are embedded is visible. It shows that some fibres have de-bonded and cleanly separated from the matrix. On other fibres matrix residue remains. The properties of SFRTP materials depend on the single constituents, the overall composition and the interfaces between the different materials. Moreover, the distribution of the constituents is relevant which is influenced by material processing. Therefore, separate sections present each of these aspects before finally a summary of current approaches for the determination of fibre distributions as well as material modelling methods for SFRTPs is given.

© Springer Fachmedien Wiesbaden GmbH, part of Springer Nature 2020
F. Dillenberger, *On the anisotropic plastic behaviour of short fibre reinforced thermoplastics and its description by phenomenological material modelling*, Mechanik, Werkstoffe und Konstruktion im Bauwesen 53, https://doi.org/10.1007/978-3-658-28199-1_2

2.1 Polymeric Matrix Materials

Overall, characteristics of SFRTPs with fibre volume fractions that are significantly lower than the bulk material volume fraction are dominated by the mechanical behaviour of the polymeric matrix. The matrix material bonds to the embedded fibres, holds them in place and stabilises them against buckling. Moreover, it achieves load transfer between fibres. The mechanical properties of polymer materials are discussed in the following. More detailed accounts on the mechanical characteristics can be found in e.g. RÖSLER et al., 2008, from which the subsequent information has been extracted unless stated otherwise.

Polymers consist of a large number of synthetically processed organic macro-molecule chains. The coupling of molecules is realised by intermolecular interactions such as van-der-Waals forces, dipole and hydrogen bridges, while the individual atoms in each molecule are bound by covalent forces. A macro-molecule chain consists of multiple repetitions of basic chemical units, the organic monomer building blocks. Monomer types contained in a macro-molecule chain may vary, in which case the resulting polymers are termed copolymers. Polymers are classified by the amount of cross-linking between individual macro-molecule chains. Figure 2.2 displays different classes of polymers schematically.

a   *b*   *c*   *d*

Figure 2.2   Illustration of polymer types: *(a)* amorphous thermoplastic with no crosslinking, *(b)* semi-crystalline thermoplastic with crystalline sections (*), *(c)* elastomer with far knit crosslinking, and *(d)* thermoset with narrow knit crosslinking.

Thermoplastics show no chemical cross-linking of chains, permitting reversible melting by heating and re-solidification by cooling without a significant change of mechanical properties, DOMININGHAUS, 2008. Hence, they can be arbitrarily shaped in the molten state, see e.g. PAPATHANASIOU et al., 1997. Macro-molecule chains of thermoplastics can assume an amorphous or semi-crystalline structure. In the amorphous case, chains are unsorted and intertwined. Mechanical properties can generally be considered to be isotropic. In the semi-crystalline case, chains show a partially regular structure of crystalline phases with parallelly aligned molecules. The orientation of crystalline phases can lead to anisotropic mechanical behaviour. In contrast to thermoplastics, elastomers and thermosets contain covalent cross-linking between amorphously arranged macro-molecules. Melting of these materials is impeded since it will usually lead to decomposition or degradation of material properties. Elastomers show rubber-like behaviour with a large elastic response, while stiffness properties are comparatively low. Thermosets include the largest amount of cross-links and show brittle mechanical behaviour.

This work investigates semi-crystalline thermoplastic polypropylene (PP) matrix materials. An overview of general mechanical properties of thermoplastics will be given in the
following.

When a sinusoidal load is applied to polymeric materials in the elastic range, they
display viscoelastic properties. The displacement reaction under sinusoidal loading will
show a phase delay on the time axis in respect to the applied load. This result can be
split into an instantaneous elastic response quantified by the storage modulus and an out-
of-phase viscous response quantified by the loss modulus. Evaluating these properties
over a wide range of temperatures reveals characteristic temperature dependencies for
different types of polymers. Figure 2.3 presents exemplary temperature dependent values
for the storage modulus of a semi-crystalline polypropylene (PP) material. In a temperature
range surrounding the glass transition temperature T_g material morphology changes from a
crystalline state to an amorphous state.

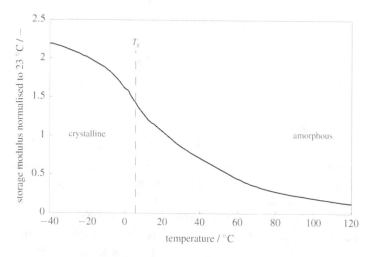

Figure 2.3 Exemplary temperature dependent storage modulus of a semi-crystalline PP.

Below the glass transition temperature, elasticity of thermoplastics is mostly related to
the stretching of intermolecular bonds such as van-der-Waals forces, dipole and hydrogen
bridges. This elastic behaviour is termed energy elasticity. Inter-molecular bonds and
macro-molecule chain segments return to their initial state upon unloading since this state
is energetically favourable. At temperatures exceeding the glass transition range, elastic
behaviour can be described by thermodynamic relations. Zones of highly entangled macro-
molecules are stretched upon loading. Higher molecule mobility at elevated temperatures
facilitates this process. The re-orientation and alignment of macro-molecules leads to a
reduction of entropy. When external loads are removed, the macro-molecules return to the
initial entangled state of higher entropy. Such an elastic behaviour is termed entropic elas-

ticity. Entropic elasticity occurs in the entangled amorphous sections of a semi-crystalline thermoplastic. Crystalline zones show energetic elastic behaviour.

When a thermoplastic material is subjected to a uniaxial load it shows a typical stress-strain response. The general stress-strain response of semi-crystalline thermoplastic for uniaxial tensile loading is given in figure 2.4.

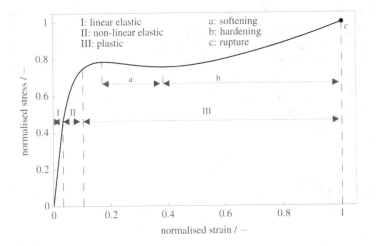

Figure 2.4   Exemplary uniaxial stress-strain response of a thermoplastic material.

The overall stress-strain response is a result of the movement within and in between entangled macro-molecules. In the case of elastic deformation, the previously described mechanisms occur. Linear stress-strain deformation in this region, as shown in figure 2.4, I, is completely reversible and, ideally, returns to the initial state without time delay. The linear elastic range may be almost non-existent for some polymers. Highly non-linear deformation is characteristic for thermoplastic materials. When the proportionality limit is passed, the stress-strain curve increasingly shows non-linear behaviour, see figure 2.4, II. Deformation is viscoelastic in this region. Inter-molecular bonds and macro-molecule chain segments are moved from their energetic equilibrium state, to which they return upon unloading with a significant time dependence. In case the movement of chains surpasses their possibility of returning to the initial state, irreversible plastic deformations arise, see figure 2.4, III. In contrast to most metals, the limit of elastic behaviour, respectively initialisation of plasticity cannot be clearly identified for thermoplastics. A clear distinction of a yield value from stress-strain curves is not possible since the transition from purely elastic behaviour to irreversible deformations is smooth. Plastic deformation will occur before the first maximum stress value is reached. During plastic deformation, relative movement between macro-molecules occurs. Since the molecules have to slip past each other, their spatial arrangement and the amount of entanglements are relevant.

Plastic deformation is usually initialised due to local stress concentrations as may occur locally for example at defects, impurities or filler phases. A stress maximum with subsequent macroscopically distinguishable softening is typical for polymeric materials and may occur under tensile, shear and compressive loads, BROWN et al., 1968. Stresses required to develop deformation decrease in the softening region, see figure 2.4, a. During plastic deformation, temperature may increase locally due to inner friction. At the initialisation of plasticity this would further promote softening of the material. However, KUNKEL, 2018 has shown that the adiabatic temperature increase is negligible at the onset of plasticity for talcum filled PP composites and softening was explained by a de-bonding of the filler and the matrix material.

Hardening will occur after a further increase of plastic strain, see figure 2.4, b. The hardening effect is a result of macro-molecule chains increasingly being stretched and orienting themselves in to the direction of stress. The depicted mechanisms may lead to necking of the specimen during tensile testing. In this case, local deformation by stretching of molecules will propagate through the sample. Macro-molecules in the fixture regions of tensile bars will remain entangled until the evolving necking zone reaches them. When fracture stress is reached covalent molecule bonds fail, causing rupture of the specimen, see figure 2.4, c. During tensile loading, elongation of chain molecules can cause the formation of cavities. This effect is termed crazing. With increasing deformation cavities or micro voids evolve, causing macroscopic volume dilatation and often leading to a colour change, referred to as stress whitening. Such an effect might also occur under biaxial tension, compare JUNGINGER, 2002.

Mechanical properties of thermoplastic materials depend strongly on loading conditions. Among these are environmental influences such as humidity, temperature and surrounding media. In case of raised temperatures, thermoplastic materials will soften and show higher ductility due to an increased mobility of macro-molecules.

Mechanical properties of semi-crystalline thermoplastics such as PP are strongly influenced by the amount of crystallinity, see MENYHÁRD et al., 2015. Depending on the distribution of crystalline phases, anisotropic mechanical properties may arise. Furthermore, the closer packing of macro-molecules in semi-crystalline sections leads to increased intermolecular bonding which results in e.g. higher elasticity and melting temperature. The degree of crystallinity may grow as part of an ageing process, as described in e.g. PONGRATZ, 2000. Introduction of reinforcement fibres leads to different crystallisation degrees and structures in comparison to a pure thermoplastic material, cf. MONEKE, 2001.

An inverse effect to temperature increase is displayed when loading velocities are increased. This will lead to a higher stress response and can provoke an embrittled material behaviour. On the macro-molecule scale, strain rate dependency can be ascribed to the restriction of inter-molecular slip because of non-instantaneous loosening of inter-looping molecule chains. Constant loads can lead to creep reactions whereas constant strains can cause load relaxation.

With thermoplastic materials the type of loading, respectively the stress state, plays a crucial role for the resulting material response. Therefore, uniaxial tensile, uniaxial compressive, shear and biaxial loading states will each result in differing mechanical behaviour. The differences concern yield strength values as well as overall stress-strain response and fracture strains. A consideration of the macro-molecule structure presented in figure 2.2 *(a)* can give an explanation for this behaviour, cf. JUNGINGER, 2002. Bundled macro-molecules will be constantly stretched and aligned and slip against each other in case of uniaxial tension, leading to the hardening seen in the stress-strain curve. Stress-strain curves during compressive loading show a form qualitatively similar to the one given in figure 2.4. In this case, softening at the beginning of plastic deformation is caused by large local deformations due to shear bands, see QUINSON et al., 1997. Compressive loading may lead to stability failure by buckling of samples, depending on the degree of test specimen slenderness, compare JERABEK et al., 2010. In case of compressive loading, hardening is a result of macro-molecule clusters being constantly packed, leading to increased surface forces acting between molecules. Shear loading will lead to a slip between intertwining macro-molecules without stretching of the molecules. Thus, almost no hardening will be observed in the quasi-static shear state.

Considering the previous statements on the mechanisms of polymer deformation, it shows that plastic deformation does in general not occur isochorically, BECKER et al., 2008. Molecule stretching and untangling during tension can lead to volume increase, while compaction under compression can lead to decreasing volumes. Phenomenologically non-isochoric deformation results in Poisson's ratios unequal to 0.5. The total amount of volumetric straining varies for different thermoplastic materials.

When thermoplastic materials are reinforced with adequately high fractions of fibres the global plastic deformation capability of the composite can appear highly decreased in comparison to the pure matrix material. Considering figure 2.4, global failure can occur shortly after region II. Nevertheless, locally the matrix material may still experience large plastic deformations.

2.2 Reinforcement Fibres

Fibres mainly account for the anisotropic properties of SFRTPs. In order to reinforce the previously discussed thermoplastic matrix materials, fibres should be of high stiffness and strength. Mostly artificial fibre materials such as glass and carbon are applied, due to their increased reinforcement capabilities based on high stiffness properties. In the context of environmentally friendly processing, an increasing usage of natural fibre materials, such as flax and hemp can be observed, cf. FU et al., 2009. Details on reinforcement fibres can be found e.g. in SCHÜRMANN, 2007 from which the following aspects have been extracted.

While showing reduced mechanical load bearing capabilities in comparison to carbon, more cost effective glass fibres are the main reinforcement material for thermoplastic materials processed by injection moulding. Mechanical properties of inorganic glass fibres are isotropic. Glass fibres show approximately linear-elastic behaviour until brittle fracture occurs without any ductile plastic deformation. Effects such as creep and relaxation or strain rate dependent behaviour are negligible in comparison to the thermoplastic matrix materials. While different glass materials are applied as reinforcement fibres, E-glass (E = electrical) is the most commonly used.

High load bearing capabilities of small fibres can be ascribed to the size effect. A single fibre of small volume will display increased strength in comparison to glass materials of larger volume. This can be explained by the weakest link theory, stating that overall strength is defined by the weakest component in a body. Strength of a brittle material such as glass is influenced by imperfections. Due to the fact that stress peaks can not be relieved by plastic deformation, fracture is induced at these imperfections. The larger a component is, the larger its imperfections may be. Moreover, the statistical probability of a higher amount of imperfections located in the load path is increased. The opposite effect occurs with reduced volumes. Hence, small fibre geometries are advantageous.

Glass fibre diameters range from 5 μm and 24 μm. Fibre reinforced composites are differentiated by the length of embedded fibres. Fibres with lengths surpassing 25 mm are designated as continuous fibres. Fibres with lower lengths are termed discontinuous. Herein constituents with lengths ranging between 1 mm and 25 mm are classified as long fibres. The short fibre classification is given for lengths from 0.1 mm up to 1 mm. Thermoplastic composites enhanced by such short fibres are the topic of this work. They are used for a multitude of technical problems and have found wide use in automotive applications, the field of civil engineering, the aerospace sector, marine applications, as well as household products and electronics, cf. FU et al., 2009.

2.3 Composite Properties

The behaviour of thermoplastics can be adjusted by adding fillers, reinforcing materials and additives, compare DOMININGHAUS, 2008. Fibre reinforcement volume fractions mainly used in case of SFRTPs in engineering applications range from 10 % up to 60 %. Since the focus of this work is set to these materials, the term composite from here on refers to SFRTPs processed by injection moulding.

Depending on geometries and orientation distributions, fibre reinforcement significantly stiffens and strengthens the original matrix material. Hardness, long term stability and thermal expansion properties are improved. However, failure strain will be reduced in comparison to non-reinforced materials. SFRTPs with an adequate amount of fibres do not show the necking behaviour previously described in section 2.1. Stress and strain distributions in SFRTPs differ from those found in continuous fibre reinforced polymers, since inhomogeneous stress concentrations arise in the matrix material at the ends of the short fibres. In general, microscopically inhomogeneous stresses and strains are generally distributed evenly throughout the samples and will lead to an overall macroscopically uniform plastic deformation of tensile bars.

Generally, the discussed phenomenological characteristics of thermoplastics will remain when they are reinforced with fibres. Among these are increased ductility, strain rate dependency, temperature dependency and stress state dependency. This is due to the fact that the matrix will still have a strong influence on the overall mechanical response. Additionally, mechanical properties of injection-moulded short fibre reinforced components are anisotropic. Such a behaviour is induced by the orientation of the embedded fibres. Anisotropic characteristics depend strongly on prevailing fibre distributions. If fibres such as transversely isotropic carbon fibres are used, the anisotropic properties of components are further pronounced.

An additional important aspect of the composite behaviour is the bonding of the fibres to the matrix material. Interface characteristics depend on the chemical properties of the constituents. In order to increase bonding, additives are added to the matrix. Moreover, the fibre surfaces are modified by the application of special coatings. Besides protecting the fibre surfaces from abrasion during processing, these coatings lead to an improved matrix melt wettability and function as coupling agents between fibres and matrix, compare SCHÜRMANN, 2007. Coupling agents have to bond with the organic polymer materials and the inorganic fibres and thereby interconnect the two constituents. The bonding interface properties strongly influence for example the fracture behaviour of composite materials, see e.g. BADER et al., 1973 and PEGORETTI et al., 1996. The influence of interfaces is discussed e.g. in SAMBALE et al., 2017.

In summary, overall mechanical properties of the composite will depend on mechanical characteristics of the constituents, geometrical features of fibres and properties of the fibre-matrix interface as well as the orientation and the distribution of fibres. The latter are strongly influenced by the processing technology explained in the next section.

2.4 Injection Moulding

Thermoplastic materials are mainly processed by injection moulding methods that allow the production of large component quantities with high cost efficiency. The injection moulding process is summarised in short herein.

In case of SFRTPs, fibres are first chopped and then compounded to the polymer matrix. Resulting compounds of fibres and matrix are stored as granulate. During the injection moulding process the fibre packed thermoplastic matrix granulate is melted, compacted and blended in a plasticizing unit, consisting mainly of a rotating screw in a heated barrel. During this procedure, fibres included in the granulate are subjected to high mechanical stresses that can lead to further reduction of the fibre length due to breakage of fibres. The screw conveys the melt to the nozzle until an adequate amount of melt is available. In the next step, the injection phase is initiated. In the first phase of an injection moulding process, namely the filling phase, the melt is injected from the nozzle into the mould cavity at high pressure and velocity. In order to apply the necessary pressure the screw acts as a ram. The melt flows into a gate from which it is distributed into the cavity. Fibres included in the melt are oriented by the melt flow during injection. Following the filling of the mould, a packing and holding pressure phase is initiated. Pressure is applied over a predefined time period, in order to compensate for the volume reduction due to the solidifying thermoplastic melt and the thermal shrinkage of the material. During a final cooling phase, the solidified thermoplastic component is cooled down until it can be extracted from the mould without deformation.

Figure 2.5 shows a cavity for the production of plate parts in an injection moulding machine. For a more complex exemplary thermoplastic part, different filling stages during the injection moulding process are displayed in figure 2.6. Further details on the process of injection moulding can be found in DOMININGHAUS, 2008 and EYERER et al., 2008 among others.

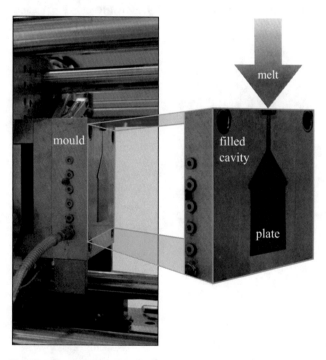

Figure 2.5 Open mould for a plate part in an injection moulding machine. The part has not yet been extracted from the cavity.

In case of complex parts, separate flow fronts may occur in the cavity, PAPATHANASIOU et al., 1997. The reason for these can be either multiple gates or part sections that lead to an interruption or split of the melt flow, such as for example bosses, ribs and holes. When the mould cavity is fully filled during the injection process, the separate flow streams may meet and merge. Mechanical properties in the resulting weld line areas are generally weaker than those of the bulk material. This property reduction especially concerns fracture strength and plastic deformation capabilities. The characteristics of weld lines are influenced by the fusing process during the packing phase of the injection moulding process and the collision angle of the respective flow fronts, cf. KUNKEL, 2018. The properties of zones where flow fronts impinge head-on will differ form those where flow fronts meet in small angles and subsequently stream through the cavity in a connected manner. The weakening effect in weld lines can be ascribed to a merely partial entanglement of the macro-molecule chains across the connecting region. In case of composites, fibres are mostly parallely aligned to the weld lines. The lack of fibres spanning the fused flow fronts will lead to a significant reduction of load bearing capabilities transverse to the weld line. Overall, weld line strength depends on the material properties, the processing conditions and the part design.

The processing of SFRTPs by injection moulding and the respective influences on the fibre orientation are depicted in e.g. PAPATHANASIOU et al., 1997, ZHENG et al., 2011 and FU et al., 2009.

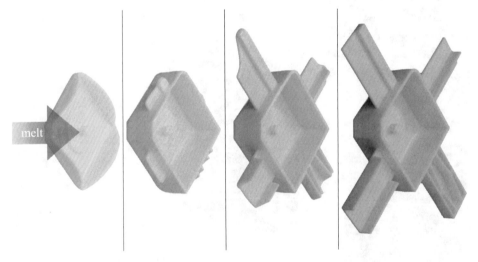

Figure 2.6   Exemplary filling stages of thermoplastic part during the injection moulding process.

A characteristic of discontinuous reinforcement is that the fibres are usually imperfectly distributed and aligned as a result of the injection moulding production process, see KLUSEMANN et al., 2009. Even seemingly simple structures such as for example rectangular plates or planar sections of the part given in figure 2.6 can show a rather complicated flow induced distribution of fibres. A typical distribution for such a part is shown in figure 2.7.

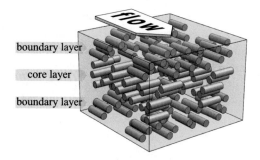

Figure 2.7   Exemplary fibre orientation in injection moulded plate structures.

Herein a short explanation of the through-thickness skin-core morphology is given. Further details can be found in PAPATHANASIOU et al., 1997. During the filling phase, after the flow leaves the gate and enters the cavity of the mould, it immediately expands. This fountain flow type filling leads to the displayed laminate structure. A lateral in-plane orientation of fibres is created in the core by the expansion, due to the fibres aligning in the extension direction that is perpendicular to the main flow direction. In this section, fibres remain in the initial orientation throughout the filling process because of a constant flow speed and a lack of shear forces. Near to the mould wall, flow is exposed to high shear due to relative velocities between core flow and melt solidifying at the mould wall. Hence, fibres are aligned by shear forces, resulting in a longitudinal orientation. Only a relatively small transitional section of reorienting fibres exists in between the longitudinal orientation near the mould wall and the lateral orientation in the centre. A thin boundary, not depicted in figure 2.7, in which a random in-plane fibre orientation can be found is located directly at the mould wall. This is a result of immediate solidification of the thermoplastic.

The mechanical material properties of thermoplastic parts depend on the filling process parameters during injection moulding and the flow properties of the thermoplastic matrix material, as researched in e.g. MÖNNICH, 2015 and KUNKEL, 2018. These parameters influence the flow conditions of the melt which lead to specific fibre orientation distributions in the composite. Distributions of fibres in injection moulded parts differ locally, since flow conditions also depend on geometrical properties of the mould. While the flow field of the polymeric melt leads to an fibre orientation distribution (FOD), additionally an fibre length distribution (FLD) is induced since fibres are often fragmented by shear forces of screws and rams during injection moulding, cf. HEGLER, 1984 and FU et al., 2009. Assuming that all fibres are of approximately equal thickness, typically merely an fibre aspect ratio distribution (FAD) is considered to characterise the distribution of fibre shapes. FODs may vary locally from random to perfectly aligned. Similarly FADs may differ locally. This generally leads to a complex process induced anisotropic micro-structural property distribution in injection moulded composite parts, regarding for example local stiffness, strength and fracture behaviour, see e.g. STOMMEL et al., 2018.

For this reason, in an finite element analysis (FEA) context, access of structural simulation models to FOD information is crucial. A simple prognosis of FODs, as can be done for laminated continuous fibre composites, cannot be performed. An integrative simulation strategy, such as schematically given in figure 2.8, permits the simulative consideration of FODs. The term integrative simulation hereby describes the linking of the results of a precursory injection moulding simulation with a structural mechanics FEA simulation, cf. HABERSTROH et al., 2006, MICHAELI et al., 2007, GLASER et al., 2008 and MÖNNICH, 2015. Each of the simulation steps requires experimental data for material model parametrisation and simulation validation.

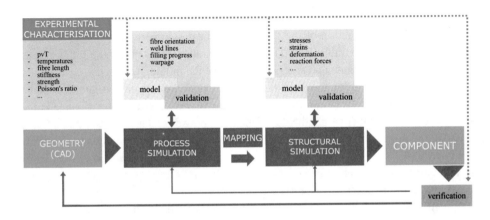

Figure 2.8   Integrative simulation process.

Injection moulding process simulation tools offer the possibility of predicting the filling process along with a calculation of fibre orientations from the velocity gradient field in the melt, considering process conditions, mould geometries and material properties, compare ZHENG et al., 2011. Up to this date, injection moulding simulation software depicts information about the fibre orientation in form of second order orientation tensors, see ZHENG et al., 2011. Details on the topic of fibre orientation tensors are presented in section 2.5. Currently, FADs cannot be predicted and only average values of the fibre length are considered. Adequate fibre distribution predictions by means of injection moulding simulation are an ongoing topic of research. Fibre orientation calculations are highly sensitive towards parameters in the rheological material description and the fibre orientation model. A poor fibre orientation prediction accuracy has a direct influence on the quality of subsequent structural simulations. Details on the topic of injection moulding simulations along with improved fibre orientation predictions were analysed in e.g. MÖNNICH, 2015. The necessity of an improved orientation information is discussed for example in MÜLLER, 2015.

FOD information is transferred to the structural simulation by a mapping process. During this process the orientation tensor along with values for the fibre volume fraction and the fibre length are transferred to the structural simulation model. A more detailed account on the description of fibre distributions is given in the following section.

2.5 Determination of Fibre Distributions

The fibre orientation plays a crucial role for the behaviour of composites. For this reason, the knowledge of the actual FOD is of major importance for the analysis of the mechanical composite properties.

FODs of materials can be described mathematically by an orientation distribution function (ODF) $\Psi(\vec{p})$ which yields a probability value for the occurrence of fibres in any orientation specified. The orientation of a single fibre f is determined by its orientation vector $\vec{p}$,

$$\vec{p} = \left\{ \cos\phi \sin\theta, \ \sin\phi\, sin\theta, \ \cos\theta \right\}^{\mathrm{T}}. \tag{2.1}$$

The angle ϕ in expression (2.1) describes the planar orientation of the fibre, while the angle θ accounts for the spatial orientation, see figure 2.9.

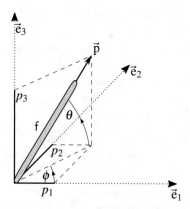

Figure 2.9 Single fibre orientation.

In numerical applications, the complex nature of ODFs make them impractical to use. Therefore, ADVANI et al., 1987 introduced fibre orientation tensors. These facilitate the numerical handling of the orientation distribution. The definitions for an orientation tensor of second order $\mathbf{A}$ and fourth order $\mathbb{A}$ are given as

$$\mathbf{A} = \oint \vec{p} \otimes \vec{p}\, \Psi(\vec{p})\, d\vec{p} \tag{2.2}$$

and

$$\mathbf{A} = \oint \vec{p} \otimes \vec{p} \otimes \vec{p} \otimes \vec{p} \, \Psi(\vec{p}) \, d\vec{p}. \tag{2.3}$$

With the symmetry condition

$$\Psi(\vec{p}) = \Psi(-\vec{p}) \tag{2.4}$$

and the normalisation condition

$$\oint \Psi(\vec{p}) \, d\vec{p} = 1. \tag{2.5}$$

ADVANI et al., 1987 show that the ODF can be fully recovered by an infinite series of even order orientation tensors given in a dyadic product formulation. Thus, an ODF can be represented more accurately if higher order orientation tensors are known. Hence, the second and fourth order orientation tensors can be interpreted as an approximation of the ODF. This becomes clear if a theoretical orientation tensor of first order is considered by neglecting the symmetry condition (2.4) in expression (2.2), which would lead to an averaged orientation vector. The second order momentum formulation of the second order tensor thus improves on this approximation and can be interpreted as the inertia tensor on the unit sphere surface, compare MLEKUSCH, 1999. This implicates that the second order tensor lacks information of higher order tensors. It can e.g. not differentiate between an FOD of equal fibre amounts aligned in two orthogonal axes and a fully random FOD, see MÖNNICH, 2015. MÜLLER, 2015 gives a detailed discussion on the validity of describing the micro-structure of a composite with transversely isotropic symmetry using only the second-order orientation tensor. Nevertheless, the second order tensor $\mathbf{A}$ is widely used in practical applications due to its simple structure. As stated in section 2.4, it is computed in injection moulding simulation software for the approximation of FODs.

The fourth order tensor, on the other hand, is not readily available. However, it is necessary for mechanical predictions in the context of integrative simulation. The missing information must be approximated by use of closure functions which commonly attempt to express the fourth order tensor by components of the second order tensor, ADVANI et al., 1987. DRAY et al., 2007 and REGNIER et al., 2008 compare multiple closure approximations. These involve the original linear and quadratic formulations by ADVANI et al., 1987 as well as more sophisticated models. Best results were achieved by the orthotropic fitted closure, which was derived by CINTRA et al., 1995. It has found wide use in engineering practice and scientific research, see e.g. KAISER, 2013. AGBOOLA et al., 2012 have shown that orthotropic fitted closures and the so called fast exact closures, more recently presented by MONTGOMERY-SMITH et al., 2011, are of similar accuracy. CHUNG et al., 2001 proposed minor improvements to the original orthotropic fitted closure formulation.

Considering the second order orientation tensor **A**, eigenvectors $\vec{q}_i$ and eigenvalues λ_i (with $i = 1, 2, 3$) can be found by solving the eigenvalue problem. These two characteristics completely characterise the tensor. Upon inspection of the tensor in the principal-axis system, defined by the eigenvectors, all off-diagonal components are zero, while the diagonal elements equal the eigenvalues. Using these findings, the second order tensor **A** can be graphically represented by an ellipsoid in a reference coordinate system, cf. ZHENG et al., 2011 and CHAVES, 2013. As shown in figure 2.10, the ellipsoid orientation is herein defined by the eigenvectors, while the magnitudes of its principal axes are determined by the eigenvalues.

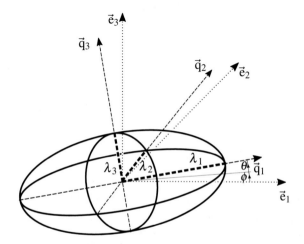

Figure 2.10   Ellipsoidal representation of the fibre orientation tensor of second order.

In order to relate the experimentally determined mechanical behaviour of the composite parts to specific FODs, the fibre distributions need to be analysed. Originally, the manual examination of composites by optical reflective microscopy was proposed. In this conventional analysis method, FODs are examined by evaluating ellipse-shaped fibre cross sections on polished micrograph sections, details on which can be found e.g. in BAY et al., 1992 and FISCHER et al., 1988. Whereas FLDs can be determined by combustion of the thermoplastic matrix and microscopic inspection of fibre lengths.

In recent years, x-ray penetration measurements by means of micro computed tomography (μCT) are increasingly adopted for the evaluation of constituent distributions, cf. BERNASCONI et al., 2012, SHEN et al., 2004 and GLÖCKNER et al., 2016. Varying densities of the constituents in a composite lead to constituent related absorption of X-rays. Therefore, the contrast between the fibres and the matrix can be evaluated and a non-destructive analysis of intrinsic fibre distributions is possible. Image analysis techniques based on filtering have been presented for the evaluation of fibres from μCT images, e.g. PFEIFER et al., 2008. The ellipse evaluation method proposed for micrograph sections has also been applied to the analysis of virtual cross sections extracted from µCT images, see GÜNZEL, 2013. GLÖCKNER et al., 2016 proposed a novel, iterative, model-based Monte-Carlo analysis percolation algorithm for the automatic recognition of fibres from µCT images. This method has significant advantages over the conventional analysis of FODs and FLDs due to reduced efforts in sample preparation and the evaluation strategy. The resulting data provided by the algorithm consists of the spatial point of origin, the orientation vector, the length and the radius of every detected fibre. A more detailed overview of different fibre orientation evaluation procedures can be found e.g. in MÖNNICH, 2015.

In an experimental context the integrals in equation (2.2) and (2.3) can be replaced by the average of a finite sum over the total number N of all found fibres, e.g. LASPALAS et al., 2007 and MÜLLER et al., 2015. This leads to the expressions

$$\mathbf{A} = \frac{1}{\sum_{f=1}^{N} w^f} \sum_{f=1}^{N} \vec{p}^f \otimes \vec{p}^f w^f \tag{2.6}$$

and

$$\mathbb{A} = \frac{1}{\sum_{f=1}^{N} w^f} \sum_{f=1}^{N} \vec{p}^f \otimes \vec{p}^f \otimes \vec{p}^f \otimes \vec{p}^f w^f \tag{2.7}$$

for experimentally derived orientation tensors.

The authors of e.g. ADVANI et al., 1987, LASPALAS et al., 2007, and CAMACHO et al., 1990 propose a value 1 for the weighting factor w^f, since they lacked information on exact geometries of fibres of a specific orientation. Later contributions suggest to take into account actual geometries and weight each fibre by its volume, see e.g. KAISER, 2013, MÖNNICH, 2015, GLÖCKNER, 2013 and MÜLLER et al., 2015. This was introduced by GLÖCKNER et al., 2016 to consider variations in radius and length, since the novel fibre analysis method allows the correlation of the fibre orientation and the fibre geometry.

The geometrical properties of the fibres are specified by the aspect ratio

$$\varphi = \frac{l}{r}. \tag{2.8}$$

The average aspect ratio of the total number N of all fibres f which are present in a specific volume section can be evaluated during experiments by

$$\varphi = \frac{1}{N} \sum_{f=1}^{N} \frac{l^f}{r^f}. \tag{2.9}$$

In this case, a weighting factor of one is applied to the determination of the average sum.

Evaluation of this data on the structural simulation level depends on the used material models. These are discussed in the following section.

2.6 State of the Art of SFRP Material Modelling

This section gives a state of the art review considering modelling of SFRTPs. Detailed accounts on the relevant modelling aspects for this work are presented in chapter 5.

The effective composite behaviour depends on mechanical properties of the embedded fibres as well as characteristics of the thermoplastic matrix material. Moreover, the spatial distribution, the orientation and the geometry of fibre inclusions influence the overall mechanical properties. These aspects must be considered in the design of components and the corresponding FEA methods in order to adequately predict anisotropic mechanical behaviour of fibre-reinforced thermoplastics. As a result, numerical simulations in the context of FEA of SFRTP materials pose numerous challenges concerning the overall simulation approach and require demanding material models. The material models have to be capable of predicting the fibre induced anisotropic mechanical behaviour. Additionally, they have to incorporate mechanical characteristics effected by the matrix material, such as non-linear plastic behaviour, stress state and strain rate dependency. Composite material models found in available FEA simulation tools such as LS-DYNA, Abaqus and others have mostly been developed for continuous fibre applications and are based on linear behaviour. This makes them infeasible for SFRTP materials which show a strongly non-linear material response.

Multiple SFRTP modelling approaches can be found in related literature. The available methods can generally be classified into two groups. These will be termed the micro-mechanical approach and the phenomenological approach in this work.

The micro-mechanical modelling techniques account for the constituents on a micro-mechanical level by individually evaluating the fibre and the matrix behaviour along with their fractional physical composition in order to predict overall performance of an ideal composite, GROSS et al., 2011. Relevant material properties include the volume fraction, the FLD, as well as the mechanical characteristics of the fibres and the matrix material. Most of the micro-mechanical models have been introduced for the prediction of unidirectional (UD) composites. Detailed accounts on established and validated micro-mechanical methods feasible for glass fibre reinforced thermoplastics can be found in MURA, 1982, TUCKER et al., 1999, KLUSEMANN et al., 2009, GROSS et al., 2011 and BÖHM, 2017 amongst others. A summarised overview is given in the following.

An simple algebraic micro-mechanical model is presented by the shear lag theory of COX, 1952. Herein an UD composite is modelled by an assemblage of parallelly arranged fibre cylinders surrounded by a matrix. Fibres are considered to account for axial load bearing while the matrix is assumed to transfer solely shear forces. Several researchers have adapted and extended the shear lag theory, e.g. HASHIN et al., 1964. A major drawback of shear lag based theories is that, as a result of the considerable model simplifications, several parameters are not readily physically graspable. In most cases, parameters, such as the critical fibre length, are iteratively adjusted to fit the model response to experimental data, e.g. LASPALAS et al., 2007. The shear lag theory additionally applies the simple micro-mechanical Rule of Mixtures approach that is also used in classical laminate theory considered for continuous fibre modelling, cf. SCHÜRMANN, 2007. It is based on the Voigt and the Reuss bound estimations of purely parallelly or purely serially aligned fibre-matrix sections for the calculation of longitudinal and transverse mechanical properties. KELLY et al., 1965 presented a model based on a modified Rule of Mixtures combined with the shear lag theory. These Kelly-Tyson expressions were adapted by FU et al., 1996 in order to predict the strength of a discontinuous SFRTP composite. Due to their simple mathematical formulation the Kelly-Tyson equations are widely used for strength predictions of SFRTP, see e.g. LASPALAS et al., 2007, BERNASCONI et al., 2011 and SAWADA et al., 2017. The well established Halpin-Tsai model offers semi-empirical expressions for the prediction of SFRTP stiffness, HALPIN et al., 1976. It is based on a Rule of Mixtures coupled with a self-consistent approach. However, the reduced predictive quality of the aforementioned methods makes them only suitable for a rough estimation of linear elasticity and strength in simple load cases, see e.g. AMBERG, 2004.

The more complex analytical mean field theories express the effective composite properties on the macro-scale as functions of volume averages of constituent phases on the micro-scale, BÖHM, 2017. Coupling of volume averages is handled by phase concentration tensors, HILL, 1963. The multitude of models are based on the Eshelby solution for the heterogeneous strain field of an elliptical inclusion in an infinite isotropic matrix medium, ESHELBY, 1957. In this sense short fibres are approximated as ellipsoidal geometries and an ideal fibre-matrix adhesion is assumed, see TUCKER et al., 1999. Since interactions between inclusions are not considered, the Eshelby theory is limited to composites with small fibre volume fractions. An expansion of the Eshelby solution to the problem of multiple inclusions interacting with one another is presented by the explicit Mori-Tanaka method, cf. MORI et al., 1973 and BENVENISTE, 1987, as well as the implicit self-consistent approach, cf. HILL, 1965a and HILL, 1965b. While the Mori-Tanaka approach can be interpreted as the consideration of individual fibres being subjected to a surrounding matrix with a far field strain equal to the average matrix strain, fibres in the self-consistent approach are evaluated as being embedded in the effective composite medium. Especially the Mori-Tanaka method has found wide use for the prediction of SFRTP properties. It offers a time-efficient way to compute effective properties with appropriate accuracy, compare TUCKER et al., 1999. TANDON et al., 1984 derived the analytical explicit Tandon-Weng equations from the Mori-Tanaka model in order to explicitly calculate the stiffness values of composites with isotropic reinforcement fibres. These equations have found widespread use in engineering practice. Many researchers base their elastic stiffness predictions on the Tandon-Weng equations, e.g. KAISER, 2013. QUI et al., 1990 offered a simplified derivation for the stiffness properties of composites with transversely isotropic filler fibres. Nevertheless, these Qui-Weng expressions are still relatively rarely used, since, being derived for multiple inclusions, they lack the simple form of the equations for isotropic fibre inclusions by

TANDON et al., 1984. Mean field theories have been adapted to plastic material behaviour, see e.g. DOGHRI et al., 2006, KAMMOUN et al., 2011 and KAISER, 2013. However, the underlying Eshelby concept only has an exact analytical solution in the linear elastic case. In the plastic case, the Eshelby problem has to be solved numerically and non-linear material behaviour has to be linearised for example by considering a plastic secant modulus, PIERARD, 2006. Generally, the prediction of mechanical properties in the plastic range is not as accurate, as the prognosis of elastic response and expressions have to be adjusted in order to reduce overly stiff plasticity approximations, cf. DOGHRI et al., 2006, PIERARD et al., 2007 and KAISER, 2013. This is due to the fact that the mean field approach only considers average stress-strain fields within constituent phases. Fluctuations and peaks are neglected that especially become relevant during plasticity for example at the fibre edges.

Apart from mean field micro-mechanical models based on analytical expressions, some authors propose a fully detailed numerical FEA of a representative volume element cell for the effective composite property prediction. The fibres and the surrounding matrix are modelled by a detailed finite element mesh, and each of the constituents is assigned an adequate material model in order to compute the heterogeneous material behaviour, cf. e.g. GUSEV, 1997, HINE et al., 2002, LUSTI, 2003, KAISER, 2013 and MÜLLER, 2015. The macroscopic behaviour of the composite components can then be modelled by transferring the calculated material parameters to appropriate phenomenological material models. Nonetheless, these methods have only found seldom use in engineering practice due to high computational costs and are therefore not considered in this work.

Most of the micro-mechanical models are based on the evaluation of UD fibres. For this reason, methods have been derived in order to link these properties to the effective behaviour of SFRTPs with random FODs of arbitrary degrees. Works such as QUI et al., 1990 have derived methods to directly take into account FODs in their equations adapted from the Mori-Tanaka model. It has been shown by researchers that multi-stage homogenisation approaches lead to significantly better results than the direct methods, BENVENISTE, 1987. These multi-stage approaches consider the fibres to be uniaxial in a first homogenisation step and take into account FOD in a second homogenisation step. This so called orientation averaging method is capable of mapping the uniaxial parameters to an oriented state by basically rotating the uniaxial stiffness matrix according to the fibre orientation distribution function, cf. ADVANI et al., 1987. Essentially, the second homogenisation step interprets the SFRTP as a meso-structure being composed of multiple UD grains, see BÖHM, 2017. Therefore, the UD configuration of the first homogenisation step is often termed pseudo grain, compare DOGHRI et al., 2006. Handling of FODs was computationally improved by the introduction of the orientation tensor, as discussed in section 2.5. The orientation averaging approach has been validated with experiments and has found wide appreciation from researchers that rely on the Mori-Tanaka model, e.g. CAMACHO et al., 1990, DOGHRI et al., 2006, ADAM et al., 2009, KAMMOUN et al., 2011, AMBERG et al., 2013 and KAISER et al., 2014. However, some works that make use of different modelling techniques such as the self-consistent approach have shown that reasonable results can even be achieved by the direct homogenisation method, see MÜLLER, 2015.

While micro-mechanical methods make the effort to describe the overall composite behaviour by a more detailed focus on the description of the intrinsic constituents, they nevertheless still represent a blurred view of the actual composite. This becomes clear upon inspection of the SEM image in figure 2.1. It is evident that for example complex interactions of fibres and fibre-matrix-interface properties cannot be completely represented by the aforementioned micro-mechanical approaches. Accordingly, the models are always based on an idealisation of the actual mechanical characteristics on the evaluated scale.

The second group of material models for SFRTP can be classified as phenomenological approaches. Phenomenological modelling methods describe the overall composite behaviour as a whole without specifically considering the individual constituents. They evaluate anisotropic fibre influences on the composite macro level and therefore present a blurred modelling technique irrespective of the exact composition. Anisotropy is hereby considered in varying degrees of detail.

First of all, this involves the greatly simplified isotropic treatment of material behaviour. Experiments in specific loading directions or averages of mechanical reactions from different loading directions are linked to an isotropic material model, compare SCHÖPFER, 2011. This approach can be considered current engineering practice in preliminary design phases, but it is even used in detailed design phases for e.g. crash applications due to robust, easy to use and computationally efficient material models, see GRUBER et al., 2011. Material models originally developed for the simulation of isotropic thermoplastics are used in this context. Since anisotropic effects are neglected experimental investigation for the derivation of material parameters can be generally kept to a minimum. The disadvantages of this concept are evident as the fibre induced orientation dependency is completely neglected. This may lead to imprecise predictions when pronounced fibre orientations are present, cf. SCHÖPFER, 2011. It can still lead to reasonable results in components under specific loading conditions with complex fibre distributions that macroscopically resemble an isotropic configuration, as is shown in SCHÖPFER, 2011.

More advanced phenomenological approaches comprise the consideration of directional dependency, by mapping material behaviour from different load directions on transversely isotropic or orthotropic material models. In this context, all material models are taken into consideration that allow the representation of a non-linear, anisotropic and strain-rate dependent stress-strain behaviour and preferably include a directional and strain-rate dependent failure criterion. Generally, the use of well established material laws originally developed for the modelling of anisotropic rolled steels are proposed, which are readily available in most FEA codes such as e.g. the software LS-DYNA, compare e.g. JENNRICH et al., 2014. However, these models mostly do not include all the material characteristics mentioned above. For this reason, some works propose the combination of multiple material models in a multi-layered virtual laminate structure, see e.g. NUTINI, 2010, KOLLING et al.,

2010, SCHÖPFER, 2011, GRUBER et al., 2011, JENNRICH et al., 2014 and WITZGALL et al., 2015. The term virtual clarifies that laminate structure does not represent the actual physical material structure, but is merely used to realise the coupling of multiple material models in order to predict SFRTP behaviour. Validation experiments in the respective papers show reasonable agreement with simulation results.

In order to adequately represent mechanical characteristics of SFRTP behaviour, several authors proposed novel phenomenological material models. For a long glass fibre reinforced PP KOUKAL, 2014 presented an elasto-viscoplastic model with a yield surface based on Hill's model. The consideration of different yield stresses under compression and tension is realised by an initial kinematic hardening of the yield surface. In KRIVACHY, 2007 and KRIVACHY et al., 2008 a material model was introduced for SFRTP using a modified piecewise yield surface based on Hill's orthotropic yield formulation to account for pressure dependency. VOGLER, 2014 proposed a more advanced model that is based on an invariant formulation and takes into account characteristics such as non-isochoric plastic deformation, anisotropic elasticity and plasticity with strain rate dependency, damage considerations and an anisotropic failure surface. The constitutive equations presented in VOGLER, 2014 are based on the use of structural tensors which allow a coordinate system free representation of the anisotropic material laws. An overview of the field of structural tensors is given for example in SCHRÖDER et al., 2010. In HOFFARTH, 2016 and HOFFARTH et al., 2017 an alternative formulation for a material model of similar capability was introduced regarding orthotropic elasticity and plasticity with pressure and strain rate dependency, damage and orthotropic failure. It is based on a Tsai-Wu yield surface and integrated into LS-DYNA. Originally, it was developed for continuous fibre composites but due to its non-linear capabilities it can also be applied to SFRTPs. DEAN et al., 2018 suggested an anisotropic finite strain material model based on the multiplicative elasto-plastic decomposition of the deformation gradient by applying invariant formulations. Overall, the presented phenomenological models require a complex parametrisation process based on a wide array of experiments.

It should be mentioned that, in the context of deriving a simplified approach at an intermediate scale between complex fully-coupled micro-mechanical models and purely phenomenological description, a material model was suggested by NOTTA-CUVIER et al., 2013. This alternative modelling strategy regards the composite as an assembly of a matrix medium and multiple fibre media representing different orientations which are linked by a multiplicative decomposition of the deformation gradient. Fibres are considered to carry loads only in their axis orientations and show one-dimensional linear elastic behaviour. The considered matrix medium can carry loads in all directions and can be modelled by arbitrary combinations of yield and hardening formulations. Specifically a Drucker-Prager type yield formulation was considered. The overall anisotropic elasto-plastic composite behaviour is modelled by an additive composition of the plastic potential. While the originally suggested model showed reasonable composite predictions in transversely directions it behaved poorly in flow direction. The model by NOTTA-CUVIER et al., 2013 has been recently extended to viscoelastic viscoplastic behaviour including damage and fracture by NCIRI, 2017. It has been shown to adequately predict the behaviour of polypropylene reinforced with 30 wt.% and 40 wt.% short glass fibres. However, this approach is not based on classical micro-mechanical models.

Most of the phenomenological models found in relevant literature, cannot directly take into account different FODs intrinsically. They do not consider the full information of an orientation tensor comprising of a main orientation and an anisotropy degree. Especially the rolled steel based or continuous fibre based material models only take into account main material orientation, which in most cases does not sufficiently describe the mechanical characteristics of SFRTPs. The aspect of different anisotropy degrees is mostly neglected or handled in an experience based manual or by an automated mapping process, see e.g. NUTINI, 2010, SCHÖPFER, 2011, JENNRICH et al., 2014 and CONVERSE, 2018. This involves the definition of locally differing material models. The respective FOD dependent material parameters have to be evaluated prior to using the material model for every FOD that needs to be taken into account. In CONVERSE, 2018 it is proposed to scale the yield ratios of a Hill criterion as a function of the orientation tensor.

In order to include the micro-mechanical composite evaluation capabilities, most FEA software requires the coupling with costly additional software tools such as Digimat, see DOGHRI et al., 2006 and DIGIMAT, 2006. Only recently, mean field micro-mechanical material models are being included into commercial FE codes such as LS-DYNA, REITHOFER et al., 2018. Alternative concepts map precomputed material data to the FEA software, see CONVERSE, 2018. When non-linear material behaviour is considered, fully micro-mechanical based codes are computationally costly. Due to numerical complexity mostly simplified yield surface models are implemented for the thermoplastic phase, compare KAISER, 2013. Micro-mechanical material models are mostly not used in the context of large simulations in which computation time is crucial such as e.g. crash applications. Advantages of micro-mechanical modelling techniques are their rather simple parametrisation and low experimental effort for the derivation of modelling parameters. FOD information needed by micro-mechanical models for structural simulations can be directly derived from preceding injection moulding simulations. The fibre orientation tensors computed by these simulations are assigned to each integration point of the finite element mesh, where they are accessible to the micro-mechanical calculations, STOMMEL et al., 2018. In the evaluation of plastic material response, phenomenological approaches offer computational advantages over micro-mechanical approaches, since they require less iterative calculations. Considering parametrisation aspects, the advanced phenomenological approaches are mostly disadvantageous to the micro-mechanical methods since parametrisation and the required experimentation can be extensive for complex components. This is due to the fact that an appropriate material model has to be defined for every local FOD. Accurate determination of material parameters thus requires a large number of experiments in different loading directions for different local FODs. As a result, many researchers use micro-mechanical techniques to at least derive elastic parameters for phenomenological methods, but there seems to be no consensus on the derivation of plastic parameters.

As a conclusion a material description that is capable of combining the versatility and simple parametrisation of micro-mechanical models with computational advantages of phenomenological methods would be desirable. Based on a thorough experimental evaluation of SFRTP behaviour, such a material model is presented in this work.

3 Mechanical Basics

The following sections give an overview of basic mechanical relations. These are needed for the evaluation of the experimental results in chapter 4 and the derivation of the material model in chapter 5. Section 3.1 gives a summary of strain related quantities. It is succeeded by a description of stress values in section 3.2. Constitutive relations are necessary in order to couple the strains and stresses. The relevant material laws are presented for elasticity in section 3.3 and for plastic behaviour as well as damage assessment in sections 3.4 - 3.7.

Details on the aspects considered in this chapter can be found in textbooks such as e.g. SIMO et al., 1998, DUNNE et al., 2006, SOUZA NETO et al., 2011, BORST et al., 2012 and ALTENBACH, 2015 as well the references given in the specific sections. The following summarised accounts were extracted from these references.

3.1 Kinematics

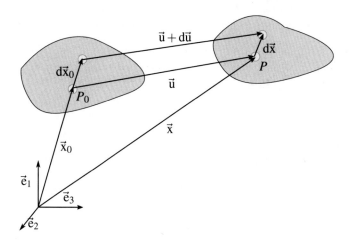

Figure 3.1 Relations between initial and current configuration of a continuum body.

In the three-dimensional Euclidean space any point P on a continuum body can be described by its position vector $\vec{x}$ in an appropriate reference system. In figure 3.1 a global right handed Cartesian base of arbitrary origin which is spanned by three orthonormal vectors $(\vec{e}_1, \vec{e}_2, \vec{e}_3)$, is defined as reference system. Such a global reference system is

© Springer Fachmedien Wiesbaden GmbH, part of Springer Nature 2020
F. Dillenberger, *On the anisotropic plastic behaviour of short fibre reinforced thermoplastics and its description by phenomenological material modelling*, Mechanik, Werkstoffe und Konstruktion im Bauwesen 53, https://doi.org/10.1007/978-3-658-28199-1_3

termed material coordinate system. Characterisation of body motion with respect to material coordinates is referred to as Lagrangian description. Deviation of a point from the initial configuration can occur by rigid body translation, rigid body rotation and stretch or distortion of the continuum. A deformation is termed rigid if all distances between arbitrary points of a continuum are preserved. The current configuration state of a point on a deformed body is given by its position vector $\vec{x}$. The displacement vector $\vec{u}$ describes the translation of point P_0 from an initial position given by $\vec{x}_0$ to a point P with the current position determined by vector $\vec{x}$ by the relation

$$\vec{u} = \vec{x} - \vec{x}_0. \tag{3.1}$$

Applying this statement, the displacement of an arbitrary point on the infinitesimal line $d\vec{x}$ in relation to the reference state $d\vec{x}_0$ is given by the expression

$$\vec{u} + d\vec{u} = (\vec{x} + d\vec{x}) - (\vec{x}_0 + d\vec{x}_0). \tag{3.2}$$

Herein $d\vec{u}$ designates the total differential

$$d\vec{u} = \frac{\partial \vec{u}}{\partial \vec{x}} d\vec{x}_0 \quad = \mathbf{H} d\vec{x}_0. \tag{3.3}$$

Relative displacement between neighbouring points is depicted by the displacement gradient $\mathbf{H}$ given in equation (3.3). The displacement gradient can be separated into a symmetric stretch tensor and an asymmetric rotation tensor, of which the second merely describes rigid body rotation.

The transformation of an infinitesimal vector $d\vec{x}_0$ to the deformed configuration $d\vec{x}$ is determined by the deformation gradient $\mathbf{F}$, as given in

$$d\vec{x} = \frac{\partial \vec{x}}{\partial \vec{x}_0} d\vec{x}_0 \quad = \mathbf{F} d\vec{x}_0. \tag{3.4}$$

It will be equal to the identity tensor in case no deformation occurs. An inverse deformation gradient

$$\mathbf{F}^{-1} = \frac{\partial \vec{x}_0}{\partial \vec{x}} \tag{3.5}$$

can be defined. $\mathbf{F}^{-1}$ linearly maps the current vector $d\vec{x}$ to the initial vector $d\vec{x}_0$.

The deformation gradient $\mathbf{F}$, similarly to the displacement gradient $\mathbf{H}$, consists of terms that are related to rigid body rotation and do not contribute to the distortion of $d\vec{x}$. By use of the second order identity tensor $\boldsymbol{\delta}$, the two gradients are interrelated by

$$\mathbf{F} = \mathbf{H} + \boldsymbol{\delta}. \tag{3.6}$$

Considering time dependency, the velocity $\dot{\vec{x}}$ of a point P can be computed from the derivation of the position vector in respect to time. In a similar fashion to the deformation gradient $\mathbf{F}$ in expression (3.4), a material time derivative $\dot{\mathbf{F}}$ is determined from the velocity vector by

$$\dot{\mathbf{F}} = \frac{\partial \dot{\vec{x}}}{\partial \vec{x}_0}. \tag{3.7}$$

The material time derivative can be expressed as

$$\dot{\mathbf{F}} = \mathbf{LF}, \tag{3.8}$$

with $\mathbf{L}$ being the spatial velocity gradient.

The velocity gradient tensor can be separated into a symmetrical part $\mathbf{L}^{\text{sym}}$, describing velocity of changing length and angles and an asymmetrical spin tensor or rate-of-rotation tensor $\mathbf{W}$, by

$$\mathbf{L} = \frac{1}{2}\left(\mathbf{L} + \mathbf{L}^{\mathrm{T}}\right) + \frac{1}{2}\left(\mathbf{L} - \mathbf{L}^{\mathrm{T}}\right) = \mathbf{L}^{\text{sym}} + \mathbf{W}. \tag{3.9}$$

The velocity gradients are applied in the context of time dependent problems for the determination of kinematic relations.

Polar decomposition allows the split of the deformation gradient $\mathbf{F}$ into a symmetric positive definite stretch tensor, describing distortion, and a rotation tensor $\mathbf{R}$, describing rigid body rotation. Either a right stretch tensor $\mathbf{U}$ or a left stretch tensor $\mathbf{V}$ are defined in this context as

$$\mathbf{F} = \mathbf{RU} = \mathbf{VR}. \tag{3.10}$$

The rotation tensor $\mathbf{R}$ is a proper orthogonal tensor and fulfils the conditions

$$\begin{aligned} \mathbf{R}^{\mathrm{T}} &= \mathbf{R}^{-1}, \\ \det(\mathbf{R}) &= 1. \end{aligned} \tag{3.11}$$

From this equation it also follows that

$$\mathbf{RR}^{\mathrm{T}} = \boldsymbol{\delta}. \tag{3.12}$$

By use of the squared right stretch tensor, the right Cauchy-Green tensor $\mathbf{C}$ is defined as

$$\mathbf{C} = \mathbf{UU} = \mathbf{F}^{\mathrm{T}}\mathbf{F}. \tag{3.13}$$

Similarly the squared left stretch tensor, leads to the left Cauchy-Green tensor

$$\mathbf{B} = \mathbf{VV} = \mathbf{FF}^\mathrm{T}. \tag{3.14}$$

In order to clarify relative amount of stretch in reference to the initial configuration, strain measures are derived from the stretch tensors. These are defined conveniently as such, that initial stretch results in an initial strain value of zero. Several different strain measures exist, among these are for example the Green-Lagrange strain, the Euler-Almansi strain and the Hencky strain, more details on which can be found in basic continuum mechanics literature, e.g. ALTENBACH, 2015. The logarithmic, true or Hencky strain of HENCKY, 1928 is defined in the material coordinate system by

$$\boldsymbol{\varepsilon} = \ln \mathbf{U} = \frac{1}{2} \ln \mathbf{C}. \tag{3.15}$$

By definition strain tensors are free of rigid body rotation and symmetric. This holds for Hencky's strain by considering that the logarithm of a symmetric second order tensor will lead to a symmetric second order tensor, cf. LU, 1998. The Hencky strain is defined by the halved logarithm of the right Cauchy-Green tensor, of which the symmetry can be proven by

$$\mathbf{C}^\mathrm{T} = \left(\mathbf{F}^\mathrm{T}\mathbf{F}\right)^\mathrm{T} = \mathbf{F}^\mathrm{T}\left(\mathbf{F}^\mathrm{T}\right)^\mathrm{T} = \mathbf{C}. \tag{3.16}$$

The true strain tensor has the form

$$\boldsymbol{\varepsilon} = \begin{bmatrix} \varepsilon_{11} & \varepsilon_{12} & \varepsilon_{13} \\ \varepsilon_{12} & \varepsilon_{22} & \varepsilon_{23} \\ \varepsilon_{13} & \varepsilon_{23} & \varepsilon_{33} \end{bmatrix}. \tag{3.17}$$

Herein the diagonal components describe normal axial strains, while the off-diagonal components refer to shear related values. In principal strain space all off-diagonal components assume a value of zero.

Due to the definition of operations on tensors, the computation of the logarithmic true strain tensor in equation (3.15) is only applicable when the right stretch tensor is in its principal axes system. This is especially relevant for the experimental characterisation of materials that are not isotropic. During experiments such as e.g. tensile tests, the axis of load application mostly coincides with the principal stress system. However, due to material anisotropies, the principal strain space must not necessarily correspond with the principal stress space. This is the case when the stiffness tensor C given in equation (3.41) has components C_{ij} (with $i = 1,2,3$ and $j = 4,5,6$) that are non-zero. The resultant coupling of normal and shear strains leads to a differently oriented principal strain space. For example off-axis tensile tests with an orthotropic material or fully anisotropic materials show this effect. However, when an orthotropic material is tested in its principal material axes, the principal stress and the principal strain system are equally oriented.

When evaluating the deformation in one of the principal strain directions, the one dimensional true strain value ε can be determined by referring to the current, deformed length of the specimen in relation to its initial reference length l_0 with

$$\varepsilon = \int_{l_0}^{l} \frac{1}{\bar{l}} \mathrm{d}\bar{l} = \ln \frac{l}{l_0}. \tag{3.18}$$

The Taylor series expansion of (3.18) at the initial length l_0 leads to

$$\varepsilon = \ln \frac{l_0}{l_0} + \frac{1}{1} \frac{(l-l_0)}{l_0} - \frac{1}{2} \frac{(l-l_0)^2}{l_0^2} + \dots. \tag{3.19}$$

Equation (3.19) can be linearised by neglecting terms of higher order. The resulting strain value

$$\varepsilon^{\mathrm{eng}} = \frac{l-l_0}{l_0} \tag{3.20}$$

is termed engineering strain. It relates the change of length to the initial length l_0 of the specimen and is widely used due to its simple evaluation in an experimental context. Equation (3.20) and (3.18) can be combined to

$$\varepsilon = \ln\left(1 + \varepsilon^{\mathrm{eng}}\right). \tag{3.21}$$

However, from the previous statements it follows that true strains cannot be directly calculated from engineering strains if these are not evaluated in the principal strain space. In this situation, the analysis of the full strain tensor is required during the experiment. This allows for the evaluation of the principal strain system and the computation of the corresponding logarithmic true strain tensor. The true strain values in the principal stress space can then be determined by tensor transformation.

True strain values are applied in this work, unless specified otherwise.

3.2 Stress

Cauchy's stress definition is considered in this work. In this context, the stress vector $\vec{t}$ is defined by

$$\vec{t} = \lim_{\Delta A \to 0} \frac{\Delta \vec{F}}{\Delta A}. \tag{3.22}$$

The stress vector is oriented in direction of the section force $\Delta\vec{F}$ and relates the local section force to the local area ΔA of a section in a deformed state, as seen in figure 3.2. The Cauchy stress tensor $\boldsymbol{\sigma}$ maps the surface normal $\vec{n}$ of the section to the stress vector by

$$\vec{t} = \boldsymbol{\sigma}\vec{n}. \tag{3.23}$$

Symmetry of the stress tensor is implied in equation (3.23). It can be determined by applying the balance of angular momentum to an infinitesimal continuum element.

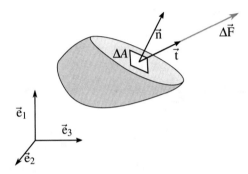

Figure 3.2   Definition of the stress vector.

The stress tensor for a general stress state is given as

$$\boldsymbol{\sigma} = \begin{bmatrix} \sigma_{11} & \sigma_{12} & \sigma_{13} \\ \sigma_{12} & \sigma_{22} & \sigma_{23} \\ \sigma_{13} & \sigma_{23} & \sigma_{33} \end{bmatrix}. \tag{3.24}$$

The diagonal stress components are related to normal axial stresses. The off-diagonal components refer to shear values.

In an one-dimensional experimental context uniaxial stress in a cross-section is calculated by relating the applied normal force F to the area of the cross-section A. In case only the initial cross-section A_0 of the undeformed body is considered, an engineering stress value σ^{eng} can be calculated by

$$\sigma^{\text{eng}} = \frac{F}{A_0}. \tag{3.25}$$

Relation of force to the actual effective cross-section of the deformed sample leads to the true stress

$$\sigma = \frac{F}{A}. \tag{3.26}$$

While the individual tensor components depend on the reference system, the overall state of stress is independent of the coordinate system. Thus, a transformation of the stress tensor may not change its general properties. This is termed the principle of objectivity or material frame-indifference. A stress tensor can be transformed to a principal axis system in which a principal stress state exists. In this case, the stress tensor is of diagonal form and contains only axial components and no shear stress values. By solving the characteristic equation

$$\det(\boldsymbol{\sigma} - \lambda \boldsymbol{\delta}) = 0., \tag{3.27}$$

the eigenvectors $\vec{q}_i$ and eigenvalues λ_i (with $i = 1, 2, 3$) of the stress tensor can be determined, cf. ALTENBACH, 2015, which fulfil

$$\boldsymbol{\sigma}\vec{q} = \lambda \vec{q}. \tag{3.28}$$

A symmetrical tensor, such as $\boldsymbol{\sigma}$, has at least three orthogonal principal directions and at most three real eigenvalues. The eigenvectors coincide with the principal directions, while the eigenvalues are the principal stress values in the respective directions. From the solution of (3.27) it can be seen that tensor invariants exists which fulfil

$$\lambda^3 - I_1\lambda^2 + I_2\lambda - I_3 = 0. \tag{3.29}$$

The principal invariants shown in (3.29) remain unchanged upon tensor transformation. They are given as

$$
\begin{aligned}
I_1(\boldsymbol{\sigma}) &= \mathrm{tr}(\boldsymbol{\sigma}), \\
I_2(\boldsymbol{\sigma}) &= \frac{1}{2}\mathrm{tr}(\boldsymbol{\sigma})^2 - \frac{1}{2}\mathrm{tr}(\boldsymbol{\sigma}^2) \quad \text{and} \\
I_3(\boldsymbol{\sigma}) &= \det(\boldsymbol{\sigma}).
\end{aligned} \tag{3.30}
$$

The hydrostatic pressure p is defined as the average value of the negative trace of the stress tensor.

$$p = -\frac{\mathrm{tr}(\boldsymbol{\sigma})}{3} = -\frac{I_1(\boldsymbol{\sigma})}{3} = -\frac{\sigma_1 + \sigma_2 + \sigma_3}{3}. \tag{3.31}$$

By extracting the hydrostatic state from the Cauchy stress tensor a deviatoric stress tensor $\boldsymbol{\sigma}^{\mathrm{dev}}$ is derived as

$$\boldsymbol{\sigma}^{\mathrm{dev}} = \boldsymbol{\sigma} - \boldsymbol{\delta} p. \tag{3.32}$$

Herein the term δp refers to the volumetric part of the stress tensor. The invariants given in equation (3.30) can also be designated for the deviatoric stress tensor.

In order to describe stress states that deviate from the uniaxial case, equivalent stress values are defined. In this context, the definition of the equivalent von Mises stress σ^{vM} is given as

$$\sigma^{vM} = \frac{1}{\sqrt{2}} \sqrt{(\sigma_{11} - \sigma_{22})^2 + (\sigma_{22} - \sigma_{33})^2 + (\sigma_{33} - \sigma_{11})^2 + 6\left(\sigma_{23}^2 + \sigma_{12}^2 + \sigma_{13}^2\right)} \tag{3.33}$$

The von Mises stress is frequently used for the analysis of isotropic stress states.

In order to describe multiaxiality of a stress state in a specimen, often the multiaxiality degree κ is used, compare KOLUPAEV et al., 2008. It is defined by

$$\kappa = \frac{p}{\sigma^{vM}}. \tag{3.34}$$

It relates the first invariant in form of the hydrostatic pressure p to the second deviatoric invariant in form of the von Mises stress σ^{vM}. Possible values of the multiaxiality factors are given in 3.1.

Table 3.1 Multiaxiality values.

Measure of multiaxiality κ	Stress state	Description
$-2/3$	$\sigma_1 = \sigma_2 < 0, \sigma_3 = 0$	equi-biaxial compression
$-1/3$	$\sigma_1 < 0, \sigma_2 = \sigma_3 = 0$	uniaxial compression
0	$\sigma_1 = -\sigma_2, \sigma_3 = 0$	shear
$+1/3$	$\sigma_1 > 0, \sigma_2 = \sigma_3 = 0$	uniaxial tension
$+2/3$	$\sigma_1 = \sigma_2 > 0, \sigma_3 = 0$	equi-biaxial tension
$\pm\infty$	$\sigma_1 = \sigma_2 = \sigma_3$	hydrostatic stress state

3.3 Linear Elasticity

In the following, merely linear mechanical behaviour will be considered for the definition of a material law in the elastic range. In the linear elastic case stresses and strains can be linked by Hooke's law. The second order stress and strain tensors are hereby mapped to each other by the fourth order stiffness tensor $\mathbf{C}$:

$$\sigma = \mathbf{C} : \varepsilon. \tag{3.35}$$

The compliance tensor $\mathbb{S}$ is defined as the inverse of the stiffness tensor by

$$\mathbb{S} = \mathbb{C}^{-1}. \tag{3.36}$$

In the general case 81 ($3 \times 3 \times 3 \times 3$) unknown material constants would need to be determined to define the fourth order tensor $\mathbb{C}$ and describe linear material behaviour. However, the number of unknowns can be reduced due to the following symmetry considerations:

$$C_{ijkl} = C_{jikl} \quad \text{applying } \sigma_{ij} = \sigma_{ji} \text{ from the balance of angular momentum and}$$
$$C_{ijkl} = C_{klij} \quad \text{from the existence of an elastic potential,} \quad \text{with } i,j,k,l = 1,2,3. \tag{3.37}$$

Furthermore, it derives from the relations in equation (3.37) that $C_{ijkl} = C_{ijlk}$ and $C_{ijkl} = C_{ijlk}$ applies. This leads to 21 unknown quantities for the stiffness tensor in the case of general anisotropy.

The expression for a stress tensor component σ_{ij} takes the form

$$\begin{aligned}
\sigma_{ij} &= C_{ij11}\varepsilon_{11} + C_{ij12}\varepsilon_{12}C_{ij13}\varepsilon_{13} \\
&\quad + C_{ij21}\varepsilon_{21} + C_{ij22}\varepsilon_{22} + C_{ij23}\varepsilon_{13} \\
&\quad + C_{ij31}\varepsilon_{31} + C_{ij32}\varepsilon_{32} + C_{ij33}\varepsilon_{33} \\
&= C_{ij11}\varepsilon_{11} + C_{ij22}\varepsilon_{22} + C_{ij33}\varepsilon_{33} \\
&\quad + 2C_{ij12}\varepsilon_{12} + 2C_{ij13}\varepsilon_{13} + 2C_{ij23}\varepsilon_{23}, \quad \text{with } i,j = 1,2,3.
\end{aligned} \tag{3.38}$$

The symmetry conditions denoted in equation (3.37) can be applied to (3.38). Further simplification is possible by considering that the strain tensor is free of rigid body rotations and is symmetric by definition, see chapter 3.1. This can be expressed by

$$\varepsilon_{ij} = \varepsilon_{ji}, \quad \text{with } i,j = 1,2,3. \tag{3.39}$$

The application of equation (3.38) to every stress tensor component can then be described by the matrix operation

$$\begin{bmatrix} \sigma_{11} \\ \sigma_{22} \\ \sigma_{33} \\ \sigma_{23} \\ \sigma_{13} \\ \sigma_{12} \end{bmatrix} = \begin{bmatrix} C_{1111} & C_{1122} & C_{1133} & C_{1123} & C_{1113} & C_{1112} \\ C_{1122} & C_{2222} & C_{2233} & C_{2223} & C_{2213} & C_{2212} \\ C_{1133} & C_{2233} & C_{3333} & C_{3323} & C_{3313} & C_{3312} \\ C_{1123} & C_{2223} & C_{3323} & C_{2323} & C_{2313} & C_{2312} \\ C_{1113} & C_{2213} & C_{3313} & C_{2313} & C_{1313} & C_{1312} \\ C_{1112} & C_{2212} & C_{3312} & C_{2312} & C_{1312} & C_{1212} \end{bmatrix} \begin{bmatrix} \varepsilon_{11} \\ \varepsilon_{22} \\ \varepsilon_{33} \\ 2\varepsilon_{23} \\ 2\varepsilon_{13} \\ 2\varepsilon_{12} \end{bmatrix}. \tag{3.40}$$

The components in (3.40) can be re-indexed by applying the operation: $11 \to 1, 22 \to 2, 33 \to 3, 23 \to 4, 13 \to 5, 12 \to 6$. The shear related strains are interpreted as shearing angles implying the doubled value of the tensorial shear strains. This leads to

$$
\begin{bmatrix} \sigma_1 \\ \sigma_2 \\ \sigma_3 \\ \sigma_4 \\ \sigma_5 \\ \sigma_6 \end{bmatrix} = \begin{bmatrix} C_{11} & C_{12} & C_{13} & C_{14} & C_{15} & C_{16} \\ C_{12} & C_{22} & C_{23} & C_{24} & C_{25} & C_{26} \\ C_{13} & C_{23} & C_{33} & C_{34} & C_{35} & C_{36} \\ C_{14} & C_{24} & C_{34} & C_{44} & C_{45} & C_{46} \\ C_{15} & C_{25} & C_{35} & C_{45} & C_{55} & C_{56} \\ C_{16} & C_{26} & C_{36} & C_{46} & C_{56} & C_{66} \end{bmatrix} \begin{bmatrix} \varepsilon_1 \\ \varepsilon_2 \\ \varepsilon_3 \\ \varepsilon_4 \\ \varepsilon_5 \\ \varepsilon_6 \end{bmatrix} . \tag{3.41}
$$

This matrix representation of the original tensor operation is termed Voigt's notation. The fourth order tensor $\mathbf{C}$ is herein reduced to a second order (6×6) matrix $\mathbf{C}$. Voigt's notation can lead to significant simplification of tensor operation calculations. This is of considerable importance for the inversion of tensors in micro-mechanical methods. However, it has to be noted that the transformation of the Voigt form is more challenging, since basic rotation operations that are applicable to tensors cannot be applied to their matrix representation, compare TING, 1996. Since the shear component order defined by the re-indexing operation is arbitrary, it can lead to confusion when stiffness values are exchanged in Voigt's notation and the re-indexing operation is not clearly stated.

A material is termed orthotropic when three orthogonal symmetry planes exist in the continuum related to its mechanical behaviour. Nine individual parameters of the stiffness tensor remain in an orthotropic material (ortho). In this case, the compliance matrix can be expressed by

$$
S^{\text{ortho}} = \begin{bmatrix} \frac{1}{E_{11}} & \frac{-v_{21}}{E_{22}} & \frac{-v_{31}}{E_{33}} & 0 & 0 & 0 \\ \frac{-v_{12}}{E_{11}} & \frac{1}{E_{22}} & \frac{-v_{32}}{E_{33}} & 0 & 0 & 0 \\ \frac{-v_{13}}{E_{11}} & \frac{-v_{23}}{E_{22}} & \frac{1}{E_{33}} & 0 & 0 & 0 \\ 0 & 0 & 0 & \frac{1}{G_{23}} & 0 & 0 \\ 0 & 0 & 0 & 0 & \frac{1}{G_{13}} & 0 \\ 0 & 0 & 0 & 0 & 0 & \frac{1}{G_{12}} \end{bmatrix} . \tag{3.42}
$$

As shown in expression (3.42) the individual elements of the stiffness or the compliance matrix can be described by engineering constants, like Young's modulus E_{ij}, Poisson's ratio v_{ij}, and shear modulus G_{ij}, with $i, j = 1, 2, 3$. The Poisson's ratio relates lateral and longitudinal strains to one another. In this work, the first index of the Poisson's ratio defines the direction of the normal strain in direction of loading, while the second index defines the direction of the negative of the normal strain transverse to loading. The Poisson's ratio relating the elastic strain in direction 2 to the strain in loading direction 1 is e.g. defined by

$$v_{12} = -\frac{\varepsilon_2^{el}}{\varepsilon_1^{el}}. \tag{3.43}$$

The Poisson's ratios are interrelated by

$$\begin{aligned} \frac{v_{12}}{v_{21}} &= \frac{E_{11}}{E_{22}}, \\ \frac{v_{13}}{v_{31}} &= \frac{E_{11}}{E_{33}}, \\ \frac{v_{23}}{v_{32}} &= \frac{E_{22}}{E_{33}}. \end{aligned} \tag{3.44}$$

Transverse isotropy or hexagonal symmetry applies when an infinite number of symmetry planes exists orthogonally to an isotropic plane. Such is the case for e.g. carbon fibres. In this context, material behaviour is equal in all orientations that are normal to the isotropic plane. The number of unknowns in transverse isotropy reduces to five engineering constants given by

$$\begin{aligned} E_{11}, & \\ E_{22} &= E_{33}, \\ G_{12} &= G_{13}, \\ v_{12} &= v_{13}, \\ G_{23} &= \frac{E_{22}}{2(1+v_{23})}. \end{aligned} \tag{3.45}$$

The stiffness matrix is of cubic symmetry when it can be completely defined by three quantities. In this case, all normal components on the diagonal are equal. Moreover, all shear components match and all non-zero off-diagonal elements have equal values as well.

In the context of isotropy (iso), mechanical behaviour is equal in all directions. This means an infinite amount of symmetry planes exists in the material. Merely two constants need to be derived for the stiffness tensor, defined as

$$\begin{aligned} E^{iso} &= E_{11} = E_{22} = E_{33}, \\ v^{iso} &= v_{12} = v_{13} = v_{23} = \frac{E^{iso}}{2G^{iso}} - 1. \end{aligned} \tag{3.46}$$

For example, glass fibres and polymeric matrix materials show isotropic behaviour. Engineering constants for exemplary isotropic materials are given in table 3.2. A Poisson's ratio < 0.5 during tensile loading describes an increase in volume, while a value of 0.5 depicts incompressible behaviour.

Table 3.2 Possible values of engineering constants for isotropic materials under tensile loading.

Material	Young's modulus N/mm^2	Poisson's ratio −	
PP	1300 to 1800	0.3 to 0.45	Osswald et al., 2006
PA66 (dry)	2000	0.4	Schürmann, 2007
PBT	2600	0.41	Schürmann, 2007
E-glass fibres	72 000 to 73 000	0.18 to 0.26	Stommel et al., 2018

In order to visualise the anisotropic characteristics of a stiffness matrix, specific properties of such a matrix can be evaluated in different directions. For example, the engineering constants can be inspected in this context. A stiffness matrix can be rotated into arbitrary orientations by use of special transformation matrices that are derived in Ting, 1996. Alternatively, by avoiding Voigt's matrix notation and using original fourth order representation, tensor rotation operations can be applied to the stiffness tensor. The specific quantity of interest can be extracted for each of the rotated stiffness matrices by consideration of expression (3.42). If one plots the quantitative amount of the extracted value against the spatial rotation angle for a sufficient amount of orientations, a smooth three dimensional (3D) surface is generated. It shows the inspected material parameter value for all orientations. Figure 3.3 presents such a spherical projection of the Young's modulus for exemplary stiffness tensors of hexagonal, cubic and isotropic symmetry. The surfaces were normalised to their respective maximum value. The geometries of these iso-spherical Young's modulus projections are typical for the respective material symmetries.

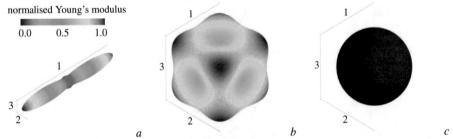

Figure 3.3 Exemplary visualisation of the iso-spherical projection of the Young's modulus for different exemplary material symmetries: *(a)* transversely isotropic, *(b)* cubic, *(c)* isotropic.

3.4 Yield Condition and Plastic Potential

In order to describe a stress state at which the elastic limit of a material is reached, a yield condition is defined. Once the yield condition is fulfilled, the region of plasticity begins and irreversible deformation of a material will occur. The definition of a yield condition Φ is given as

$$\Phi = f(\boldsymbol{\sigma}, \xi_i, \psi_i) \begin{cases} < 0 & \rightarrow \text{elastic deformation} \\ = 0 & \rightarrow \text{plastic deformation.} \end{cases} \tag{3.47}$$

To define the yield condition the stress state $\boldsymbol{\sigma}$ has to be taken into account. The boundary of the elastic region given by the yield condition can be interpreted as a surface in the six-dimensional stress domain. Thus, it is also termed yield surface. Additionally, the yield criterion depends on an arbitrary number of material parameters ξ_i and, in the case of hardening, a number of internal variables ψ_i that govern hardening. Hardening describes a transformation, respectively evolution of the yield surface in respect to the internal variables. When the yield surface is merely expanded in stress space, hardening is termed isotropic. A shift of the yield surface in stress space is termed kinematic hardening. Combinations of both hardening mechanisms may apply.

The consistency condition describes that no load points can lay outside of the enveloping yield surface. It ensures that all new plastic material states during hardening remain on the yield surface and fulfil the yield condition. The consistency condition is expressed by

$$\Phi(\boldsymbol{\sigma} + d\boldsymbol{\sigma}, \xi_i, \psi_i + d\psi_i) = \Phi(\boldsymbol{\sigma}, \xi_i, \psi_i) + d\Phi = 0. \tag{3.48}$$

In the context of small strains, elasto-plastic decomposition is applicable. In this context, total strain $\boldsymbol{\varepsilon}$ can be depicted as the sum of elastic $\boldsymbol{\varepsilon}^{\text{el}}$ and plastic strain $\boldsymbol{\varepsilon}^{\text{pl}}$ by

$$\boldsymbol{\varepsilon} = \boldsymbol{\varepsilon}^{\text{el}} + \boldsymbol{\varepsilon}^{\text{pl}}. \tag{3.49}$$

Assuming Hooke's law for the determination of the elastic strain fraction, plastic strains can be expressed by

$$\boldsymbol{\varepsilon}^{\text{pl}} = \boldsymbol{\varepsilon} - \mathbb{S} : \boldsymbol{\sigma}. \tag{3.50}$$

For the one-dimensional case equation (3.50) takes on the form

$$\varepsilon^{\text{pl}} = \varepsilon - \frac{\sigma}{E}. \tag{3.51}$$

Plastic strains arise after the yield condition is met. The flow rule couples plastic strains to plastic stresses and is given as

$$d\boldsymbol{\varepsilon}^{pl} = d\lambda \frac{\partial h}{\partial \boldsymbol{\sigma}}. \tag{3.52}$$

Herein the change of plastic strains $\boldsymbol{\varepsilon}^{pl}$ is related to the flow potential h by a scalar factor $d\lambda$ which is termed plastic multiplier. $d\boldsymbol{\varepsilon}^{pl}$ is oriented normal to the flow potential surface. In case flow potential and yield surface are equal flow is termed associative, otherwise it is termed non-associative.

In section 3.3 the definition of Poisson's ratios for elastic material behaviour was given. As shown in e.g. KOLUPAEV, 2018, plastic Poisson's ratios can be defined similarly by the relation of incremental plastic strain values. E.g. the plastic Poisson's ratio v_{12}^{pl} relating the incremental plastic strain in direction 2 to the incremental strain in loading direction 1 is defined as

$$v_{12}^{pl} = -\frac{d\varepsilon_2^{pl}}{d\varepsilon_1^{pl}}. \tag{3.53}$$

3.5 Limit Criteria

Limit or failure criteria are used to define the boundary between specific regions of material behaviour. They can be defined in terms of stress and in terms of strain. In the context of this work, on the one hand, limit criteria are used to define a transition from elastic to plastic behaviour in form of a yield condition as given in expression (3.47). On the other hand, they are used for the definition of fracture criteria, respectively for the description of maximum plastic deformation capability.

A review of isotropic yield surface formulations can be found e.g. in KOLUPAEV, 2018. The isotropic von Mises (vM) criterion is given by

$$\Phi^{vM} = \sigma^{vM} - \xi_0^{vM} = 0. \tag{3.54}$$

The von Mises criterion is based on the von Mises stress introduced in equation (3.33). It is one of the most widely used isotropic yield conditions but it is most suitable for metallic materials with a cubic crystalline structure. In principal stress space, the von Mises yield surface takes the shape of a cylinder of which the mean axis coincides with the axis of hydrostatic pressure. When being used in the context of thermoplastic materials, a major drawback of the von Mises condition is its definition in terms of the stress deviator. This leads to a negligence of hydrostatic pressure dependency, respectively volumetric deformation. As stated in chapter 2.1, mechanical behaviour of thermoplastics depends on the stress state, which can be defined in terms of hydrostatic pressure as shown in equation (3.34).

Therefore, an appropriate yield condition should take into account hydrostatic pressure p, respectively the first invariant I_1 of the stress tensor given in expression (3.30). An example of such a yield formulation is the Drucker-Prager (DP) criterion

$$\Phi^{DP} = \frac{1}{\sqrt{3}} \sigma^{vM} + I_1 \xi_1^{DP} - \xi_0^{DP} = 0. \tag{3.55}$$

The Drucker-Prager criterion has linear dependency on I_1 leading to a conical form in the principal axis system. Parameter ξ_0^{DP} is the tangent function of the angle between the yield surface and the hydrostatic axis. ξ_1^{DP} controls the intersection point of yield surface and hydrostatic axis.

The linear dependency on the hydrostatic pressure limits the application of the Drucker-Prager criterion. It will, for example, lead to un-physical, infinite limit values in multiaxial compression settings. Therefore, it has been expanded to a quadratic formulation in KOLLING et al., 2005. The semi-analytical model for polymers (SAMP) derived by the authors is actually based on the adaptation of a general anisotropic quadratic yield surface to the case of isotropy and is expressed in terms of invariant values by

$$\Phi^{SAMP} = \left(\sigma^{vM}\right)^2 - \xi_0^{SAMP} - \xi_1^{SAMP} p - \xi_2^{SAMP} (p)^2 = 0. \tag{3.56}$$

As shown in KOLLING et al., 2005 parameters ξ_0^{SAMP}, ξ_1^{SAMP} and ξ_2^{SAMP} of the yield surface given in equation (3.56) can be calculated from yield values for uniaxial tension, uniaxial compression, shear and equi-biaxial tension. The von Mises formulation can be regained by setting the parameters of the SAMP model accordingly, i.e. $\xi_1^{SAMP} = \xi_2^{SAMP} = 0$.

When anisotropic materials are considered, more sophisticated yield criteria are required. These criteria need to incorporate additional material parameters that allow the weighting of individual stress values in specific directions. Historically, anisotropic limit conditions have at first been defined for the yielding of anisotropic metal materials, e.g rolled steels in deep drawing applications, as well as for the fracture of unidirectional continuous fibre composites. Considering the possibility of describing anisotropic behaviour, and perhaps neglecting the specific physical meaning of certain parameters by only making use of their mathematical properties, some of these criteria can be adapted to short fibre reinforced composites.

A survey on anisotropic models originally derived for the modelling of yield in metal materials is given e.g. in MEUWISSEN, 1995, BANABIC, 2010 and PETERS, 2015. Some of these limit criteria have been adapted to short fibre composite materials, see e.g. NUTINI et al., 2017.

An overview of failure criteria that have been proposed for the depiction of fracture of unidirectional continuous fibre reinforced composites can be found in e.g. NAHAS, 1986 and KADDOUR et al., 2013. However, not all of the available models can be extended to arbitrary FODs, compare OSSWALD et al., 2017. Considering short fibre reinforced composites, only criteria capable of representing orthotropic behaviour should be considered, since FODs may have 3D characteristics. The simplest failure criterion applied to composites is the condition of maximum stress, see JENKIN, 1920. However, its application will mostly lead to highly over-designed components. Sophisticated anisotropic failure conditions can basically be grouped into strength tensor based models and mechanistic models.

The first group of criteria consists of generalised strength tensor based models. These models propose a single implicit expression for the determination of a scalar failure value, representing a failure surface in stress space. Herein limit normal stresses, shear stresses and in some cases interaction of these are considered by respective engineering strengths. In order to include anisotropy, the respective values are taken into account for longitudinal and lateral loads. Among these failure criteria are e.g. the Azzi-Tsai and the Tsai-Hill condition, cf. AZZI et al., 1965 and TSAI, 1965. These do not distinguish between compressive and tensile loading. The Gol'den-blat–Kopnov model by GOL'DENBLAT et al., 1965, respectively Tsai-Wu model from TSAI et al., 1971 offer the possibility of differentiating between compressive and tensile loading. A discussion of general quadratic yield criteria such as the Tsai-Wu approach is given in FENG et al., 1984. A study on this group of failure criteria can be found in SKRZYPEK et al., 2016. Piecewise extensions of tensor criteria exist, in order to allow a dissociated adjustment to experimental data in different stress quadrants, see e.g. TANG, 1989 among others.

Two examples of stress tensor based models are presented in the following. The Hill (H) yield condition, introduced by HILL, 1948, represents an anisotropic extension of the von Mises criterion and is given as

$$
\begin{aligned}
\Phi^{\mathrm{H}} = \big(& \xi_1^{\mathrm{H}}(\sigma_2 - \sigma_3)^2 + \xi_2^{\mathrm{H}}(\sigma_3 - \sigma_1)^2 \\
& + \xi_3^{\mathrm{H}}(\sigma_1 - \sigma_2)^2 + 2\xi_4^{\mathrm{H}}\sigma_4^2 + 2\xi_5^{\mathrm{H}}\sigma_5^2 + 2\xi_6^{\mathrm{H}}\sigma_6^2 \big)^{1/2} = 1.
\end{aligned}
\tag{3.57}
$$

The parameters ξ_1^{H}, ξ_2^{H}, ξ_3^{H}, ξ_4^{H}, ξ_5^{H} and ξ_6^{H} of the Hill yield formulation need to be calculated from specific yield values of the anisotropic material considered.

The Tsai-Wu criterion describes a general anisotropic quadratic yield surface, see TSAI et al., 1971. For the case of an orthotropic material, the Tsai-Wu yield condition in stress space is depicted by

$$
\begin{aligned}
\Phi = \vec{F}^{\mathrm{T}}\vec{\sigma} + \vec{\sigma}^{\mathrm{T}}\mathbf{F}\vec{\sigma} \\
= F_1\sigma_1 + F_2\sigma_2 + F_3\sigma_3 + F_{11}\sigma_1^2 + F_{22}\sigma_2^2 + F_{33}\sigma_3^2 \\
+ 2F_{12}\sigma_1\sigma_2 + 2F_{13}\sigma_1\sigma_3 + 2F_{23}\sigma_2\sigma_3 + F_{44}\sigma_4^2 + F_{55}\sigma_5^2 + F_{66}\sigma_6^2 = 1.
\end{aligned}
\tag{3.58}
$$

The terms $\vec{F}$ and $\mathbf{F}$ describe second, respectively fourth order strength tensors $\mathbf{F}$ and $\mathbb{F}$ that have been reduced by use of the Voigt notation presented in chapter 3.3. TSAI et al.,

1971 list explicit equations for the derivation of the parameters $\vec{F}$ and $\boldsymbol{F}$ from engineering strengths.

The Tsai-Wu expression has found wide use due to its simple mathematical form and manageable amount of required experiments. It allows the definition of differing yield values for different stress states such as e.g. uniaxial tension and compression. However, the Tsai-Wu criterion is often reported to overestimate material behaviour in the case of multiaxial compression, compare NEALE et al., 1989. A proper experimental derivation of the off-axis interaction terms is reportedly challenging, cf. TSAI et al., 1971.

The second group of failure criteria for composites were established to incorporate physical effects during material failure. These mechanistic models were explicitly developed for the fracture description of continuous fibre materials. They consist of multiple expressions of which each describes a specific failure type. In some models interactions between the different failure modes are considered. The Puck model by PUCK, 1996 and the Cuntze approach by CUNTZE et al., 2004 amongst others can be assigned to this family of criteria. A discussion of failure criteria for anisotropic composites can be found in CHRISTENSEN, 2013. The mechanistic models are mainly applicable to brittle laminates with thermoset matrices. The expressions could be adjusted to SFRTP materials but the original physical justification of the mechanistic criteria would by lost. This is due to the fact, that the fracture behaviour of continuous fibre reinforced materials is not comparable to SFRTP materials with arbitrary fibre orientations.

3.6 Damage

Polymers typically show a viscoelastic unloading behaviour that often exhibits reduced stiffness once plastic deformations are present. This can be ascribed to macro-molecule chains with irreversible displacements and dissolved weak covalent bonds. Additionally, in case of fibre reinforcement, failure of the fibre-matrix interface, void growth due to stress peaks and fibre fracture may continuously decrease the loading capability, compare ARIF, 2014. These effects are summarised by the term damage. Damage is often considered to occur at plastic deformations. In this context, damage initiation is defined by the material yielding point, while damage evolves throughout the development of plastic strains.

A Lemaître damage formulation will be considered in this work due to its simple form and practical applicability. Details on this, as well as alternative damage formulations, can be found e.g. in LEMAITRE et al., 2005. Damage is herein quantified by a variable d depending on plastic strain. The effective manifestation of damage is assumed to be that of the reduction of the effective cross section onto which loading is applied. This approximation stems from the assumption of a cross section reduction due to the introduction of defects, respectively voids and micro-cracks, during loading. In the case of SFRTP, these voids can be interpreted as a fracture of the matrix material in-between fibres and as a separation of polymer and fibres due to insufficient bonding. Damaged behaviour of thermoplastics can be adequately approximated by this simple damage assumption as various authors have shown, e.g. KOLLING et al., 2005, HOFFARTH, 2016 and others. The scalar damage variable d is dependent on the cross section $\tilde{A}$ related to the actual effective cross section A without any voids and damage by

$$d = 1 - \frac{A}{\tilde{A}}. \tag{3.59}$$

Hence, d is a positive, real number that takes the value zero for the undamaged state and a value of unity for the fractured state of the damaged body. The damaged cross section $\tilde{A}$ hereby refers to the total outer cross section that is measured when no knowledge of inner voids or cracks exists. Force F acting on the cross section is equal for damaged and undamaged states, see figure 3.4.

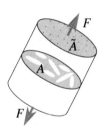

Figure 3.4   Effective and damaged cross section of a solid.

By applying the previous considerations, the effective (or undamaged) stress σ can be related to the damaged stress $\tilde{\sigma}$. This is expressed for the one dimensional case by

$$\tilde{\sigma} = \frac{F}{\tilde{A}} = \frac{\sigma A}{\tilde{A}} = \sigma(1 - d). \tag{3.60}$$

While the actual material parameter, i.e. Young's modulus E, remains unchanged in the damaged configuration, the measurable modulus $\tilde{E}$, when only the damaged cross section $\tilde{A}$ is known, is determined by the damaged stress with

$$\tilde{E} = \frac{\tilde{\sigma}}{\varepsilon^{el}} = \frac{\sigma(1 - d)}{\varepsilon^{el}} = E(1 - d). \tag{3.61}$$

Therefore, this damage description can be perceived as a stiffness softening.

It is assumed that strains are of the same magnitude, irrespective of the choice of damaged or undamaged configuration. Upon application of equation (3.51) for the one dimensional case, strain equivalence can be expressed by

$$\varepsilon^{\text{el}} = \frac{\tilde{\sigma}}{\tilde{E}} = \frac{\sigma}{E},$$
$$\varepsilon^{\text{pl}} = \varepsilon - \frac{\tilde{\sigma}}{\tilde{E}} = \varepsilon - \frac{\sigma}{E}. \tag{3.62}$$

3.7 Rate Dependency

As stated in chapter 2, the mechanical characteristics of thermoplastic materials are strain rate dependent. Elastic and plastic behaviour change at different speeds of loading. The stress response increases with growing strain rate. In the one dimensional case, viscoelastic material behaviour can be modelled by a combination of a spring and a damper. The spring with stiffness E is responsible for instantaneous reaction to loading, while the damper with viscosity η introduces a time delay. Based on the arrangement of spring and damper different models can be derived. A parallel connection leads to the Kelvin-Voigt model and a series connection leads to the Maxwell model.

In order to account for viscoplastic (vp) strain rate dependent plastic deformation, similar rheological assumptions can be made. In this case, a friction slider element is parallelly connected to the damper of the Maxwell model. This is shown for the one-dimensional case in figure 3.5.

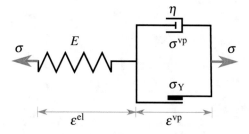

Figure 3.5 One dimensional rheological model of the elasto-viscoplastic material behaviour.

Analogous to equation (3.49), the splitting of strains can be performed for rate dependent plasticity. Total strain $\boldsymbol{\varepsilon}$ is characterised by the sum of elastic strain $\boldsymbol{\varepsilon}^{\text{el}}$ and viscoplastic strain $\boldsymbol{\varepsilon}^{\text{vp}}$ as given in

$$\boldsymbol{\varepsilon} = \boldsymbol{\varepsilon}^{\text{el}} + \boldsymbol{\varepsilon}^{\text{vp}}. \tag{3.63}$$

Regarding the parallel connection of the friction element and the damper, total stress $\boldsymbol{\sigma}$ can be depicted as the sum of the stress contributions of the friction components and the viscous fraction $\boldsymbol{\sigma}^{\mathrm{vp}}$ by

$$\boldsymbol{\sigma} = \boldsymbol{\sigma}_{\mathrm{Y}} + \boldsymbol{\sigma}^{\mathrm{vp}}. \tag{3.64}$$

Yield stress $\boldsymbol{\sigma}_{\mathrm{Y}}$ is defined as the stress at which irreversible plastic deformation occurs. Thus, the stress contribution from the friction element can be interpreted as yield stress.

It is evident that upon use of the definition given in equation (3.64) the consistency condition presented in expression (3.48) does not hold, since stresses outside of the yield envelope may occur. The viscous stress fraction is termed overstress and will return to the yield surface with a certain time delay. In BORST et al., 2012 different classes of viscoplasticity models are listed. The power law model proposed by PERZYNA, 1966 is based on the previously stated overstress consideration. In WANG et al., 1997 yield stress is interpreted as a function of rate effects ψ. This avoids the overstress assumption since rate effects are directly incorporated into the yield surface definition by

$$\Phi = f(\boldsymbol{\sigma}, \xi_i, \psi_i, \dot{\psi}_i). \tag{3.65}$$

Thus, the consistency condition found in expression (3.48) can still be applied.

4 Material Characterisation

Material testing was focussed on the characterisation of material parameters for the derivation of an adequate material model for FEA simulations. Next to the determination of material model input data, the test results were applied to the validation of simulation and homogenisation procedures. The overall experimental program consisted of mechanical tests as well as micro-structural analyses for the determination of FODs. As stated in chapter 2, thermoplastic material characteristics depend strongly on the load type and the triaxiality of the applied stress state. This is why, the analysis of the thermoplastic matrix material and the SFRTP was conducted under different loading conditions such as tension, compression, and shear. Considering strain rate dependency, the material behaviour was analysed at quasi-static strain rates smaller than $10^{-3}/s$ and multiple dynamic loads. In order to assess the material anisotropy, different loading directions were tested and the actual fibre configuration in the samples was measured.

The following sections give an overview of the conducted testing methods and experimental results.

4.1 Choice of Material

In the context of this work, a thermoplastic matrix material reinforced with short fibres was evaluated. From the broad range of thermoplastic materials a representative material was chosen in order to derive an experimental basis for material modelling. PP was selected as matrix material since it is widely used in industrial applications. Moreover, short fibre reinforced PP components are increasingly applied in automotive components, due to lightweight capabilities and comparably low material costs, see MADDAH, 2016. PP is classified as semi-crystalline thermoplastic material, as described in section 2.1. Detailed descriptions of the chemical composition and properties of this material can be found in e.g. DOMININGHAUS, 2008, EYERER et al., 2008 and BONNET, 2009. The influence of humidity on its mechanical behaviour is negligible in comparison to polyamide (PA).

In order to understand the influence of short fibre reinforcement and distinguish it from the actual matrix material behaviour, the need was identified to test a short fibre reinforced composite material and its corresponding non-reinforced matrix material side by side. Research among plastics distributors showed that none was capable of offering the exact non-reinforced matrix material corresponding to an SFRTP, due to confidentiality considering the precise constituent configuration. Hence, the decision was made to obtain a

© Springer Fachmedien Wiesbaden GmbH, part of Springer Nature 2020
F. Dillenberger, *On the anisotropic plastic behaviour of short fibre reinforced thermoplastics and its description by phenomenological material modelling*, Mechanik, Werkstoffe und Konstruktion im Bauwesen 53, https://doi.org/10.1007/978-3-658-28199-1_4

pure PP material that serves as matrix and self-compound the reinforced material needed for this work. This procedure ensured complete control of all the constituents. The compounding process was conducted at LBF. PP of type Moplen HP500N by Lyondell-Basell was chosen for this purpose. The same material batch was used for all samples. To obtain an SFRTP consisting of the same matrix, 30 wt.% short glass fibres were added to the polymeric bulk material. In order to assure stability, durability and an adequate performance of the reinforced material three additional additives were included during the compounding process:

- the long-term thermal stabiliser: 0.5 wt.% Irganox B225 by BASF,

- the calcium stearate acid scavenger: 0.07 wt.% Ceasit AV by Baerlocher,

- and the coupling agent: 3.0 wt.% Exxelor PO 1015 by Exxon Mobile.

In respect to the mass fraction of fibres this composite material is from here on referred to as PPGF30. In an effort to minimise the physical differences of the matrix material in the reinforced state, in comparison to the non-reinforced state, the aforementioned additives were compounded to the pure matrix material PP as well. Hence, in the following sections the term PP always refers to this compound of matrix and additives. Thus, apart from the glass fibres, the two material configurations PP and PPGF30 consist of the same additives, allowing an improved determination of the influence of the reinforcement fibres on overall material behaviour. This attempt of evaluating the actual matrix of the reinforced material does, however, not take into account chemical processes that may take place during the addition of fibres. Constant processing parameters where applied during the injection moulding of similar parts.

Several experimental studies concerning the mechanical behaviour of glass fibre reinforced polypropylene with similar and different fibre fractions can be found in literature. For example, a vast experimental study on the tensile properties of unidirectional PPGF30 specimens is given in Fu et al., 2000. Detailed experimental studies on the strain rate dependency of different PPGF30 materials can be found in e.g. SCHOSSIG et al., 2008, NUTINI et al., 2017 and NCIRI, 2017. In contrast to these studies, the present work derives the characteristics of the PPGF30 material alongside the actual pure PP matrix. Moreover, it additionally considers varying degrees of fibre distributions.

Micro-mechanical homogenisation methods derived in this work for isotropic glass fibres were adapted to be suitable for transversely isotropic fibres in the elastic range. For this reason, additional test results considering elastic behaviour of PP composites with transversely isotropic carbon fibres are given in the appendix. The respective materials are designated as PPCF10 and PPCF20 and are presented to show the validity of the material model for these fibre types.

4.2 Coordinate System Definition

Throughout this work extraction angles, loading angles and material orientations refer to the direction of material flow during injection moulding. The global Cartesian coordinate system is defined as such, that the y, respectively 2 axis, is oriented in direction of flow during injection moulding. On the other hand the x, respectively 1 axis is directed transversely to flow and parallelly to the surface of a component. In this context, angle designations refer to the x-y plane when not stated otherwise. Flow direction is labelled by an angle of $0°$, whereas an orientation transverse to flow is designated by an angle of $90°$. All injection moulded parts along with their associated coordinate system can be found in appendix A.

4.3 Experimental Set-up

4.3.1 Specimen Preparation

Different SFRTP components were injection moulded at LBF with the materials presented in section 4.1. All specimens needed for the experiments were extracted from these source parts. Alongside the source part geometries, the relevant samples for mechanical and microstructural testing are described in the following. Samples were not explicitly conditioned or aged. Prior to testing, they experienced prolonged exposure to constant environmental conditions of $23\,°C$ and a humidity of $50\,\%$.

4.3.1.1 Source Part Geometries

In order to evaluate geometrical influences on material behaviour, and more specifically effects of mould geometry on the FOD, different planar source parts of varying thickness and form were evaluated.

The main components were plates of the dimensions $80\,\text{mm} \times 80\,\text{mm} \times 2.5\,\text{mm}$, as seen in the appendix in figure A.1. Most of the testing was performed with samples that have been extracted from these plates. They are marked by a $\square$ symbol in this work. The triangular gating system in the inlet section, seen in figure A.1, leads to a homogeneous filling of the plate cavity with a parallel melt front during injection moulding, cf. AMBERG, 2004 and KUNKEL, 2018. As will be seen in section 4.4, these plate parts only offer a limited fibre orientation degree. Hence, additional flat bar components that have been developed by AMBERG, 2004 were produced with the PPGF30. In the relevant bar section of $80\,\text{mm} \times 20\,\text{mm} \times 2.0\,\text{mm}$, these offer an increased uniaxial directionality of fibres over the entire thickness, due to the convergent design of the gate. Samples that were extracted from flat bar source parts are depicted by a $\|$ symbol.

A detailed overview of all source part geometries and their relevant dimensions can be found in figure A.1 in the appendix.

4.3.1.2 Specimen Geometries

Source part planarity ensured that all required test samples could be extracted by mechanical milling. Figure A.2 in the appendix gives a detailed account on the positions the samples were extracted from. All specimens were extracted at an angle of 0° in reference to the flow direction. Further samples were extracted at additional angles in order to adequately evaluate the assumption of isotropy in case of the PP matrix material, respectively anisotropic characteristics due to reinforcement fibres in case of the polypropylene reinforced with 30 wt.% short glass fibres (PPGF30). Unless otherwise specified, specimens were obtained from the centre of respective source parts at position P2. Specific dimensions of all specimens used are given in figure A.3.

The geometries of tensile bar and shear specimens were analysed in detail in BECKER, 2009. In that work a gauge length of 12 mm was proposed for the tensile bar. Even though originally developed in BECKER, 2009 for non-reinforced thermoplastics under strain rate dependent loading, these geometries have since found wide use in the field of SFRTP testing, e.g. KAISER, 2013, MÜLLER, 2015, VOGLER, 2014, SCHÖPFER, 2011, STOMMEL et al., 2018 amongst others. Since these specimens have been explicitly adjusted to meet requirements of digital image correlation (DIC) measurements and high speed testing applications, they were chosen for this work.

The cubic compression specimens have been evaluated in e.g. DILLENBERGER, 2014. They fulfil the condition of being able to be extracted from the same source parts as the other specimens. This is a necessity in case of SFRTP materials with part dependent FODs. By ensuring this condition, the results from different specimens offer increased comparability, as shown in chapter 4.4. Comparison of behaviour at different stress states can be problematic if samples with different FODs are used, cf. KRIVACHY, 2007. Furthermore, the cubic samples make it possible, at least for non-reinforced polymers, to analyse a large range of plastic deformation as opposed to compression bars. This is shown for example in BECKER, 2009. By a reduced height in comparison to width, respectively thickness, buckling of samples can be avoided for non-reinforced materials.

Bending specimens were dimensioned in reference to the tensile bars. Thus, width corresponds to tension gauge width, whereas length correlates with the maximum possible extraction length from the plate.

Concerning biaxial testing, no specimen was cut from the source parts, but rather the plates were used themselves.

For an analysis of 90° behaviour of the highly oriented flat bar source part no adequate specimen existed. In AMBERG, 2004 it is suggested to insert multiple bar sections of flat bar source parts into the cavity of a compression mould, fuse the material and subsequently cool it down under pressure. This leads to unidirectionally oriented plates from which transverse specimens can be extracted. However, such specimens have multiple weld lines. While weld lines may have negligible effect on elasticity at least for a non-reinforced thermoplastic, plastic behaviour as well as rupture are strongly influenced. This is thoroughly discussed in

KUNKEL, 2018. For this reason, the alternative small tensile specimen seen in figure A.3, is proposed in this work. It is essentially a normal tensile bar that was scaled down by a factor of 0.35. This results in a gauge length of 4.2 mm. When such a sample is extracted from the relevant flat bar in transverse orientation, the possible clamping area is reduced in comparison to the non-scaled tensile bar. This is due to the fact that the flat bar width of 20 mm does not meet the scaled length of 28 mm. Multiple versions of the miniature tensile bar where tested. It was found that by a horizontal increase of the clamping area, the scaled bar can be adequately fixed in a tensile clamping system. The scaling factor 0.35 hence represents an iteratively derived compromise between an ideally large specimen in order to comprise a representative amount of fibres and a sufficient clamping area. Smaller scaling factors lead to a higher scatter of results, due to material being squeezed out of the clamp during testing. The proposed small tension bar was applied to the derivation of tensile properties of the PPGF30$^\|$ part in 0° and 90° loading directions to ensure consistent results for both loading directions.

Cylindrical samples were used for μCT analysis. μCT analysis of micro-structures requires small-sized specimens. This allows a minimal distance between x-ray source and specimen, which makes it possible to properly resolve fibre inclusions. The cylindrical form enables a simple rotational scanning of μCT samples. In order to evaluate the local variation of FOD and FLD, μCT samples were extracted at multiple positions of the source part as shown in A.2. A marking of samples prior to being extracted from the plate, allowed the FOD results to be properly located in the global coordinate system. Such a mark is seen in figure 4.11 *(a)*.

4.3.2 Determination of Mechanical Properties

Machine configurations and conditions used for mechanical testing are summarised in the following section, alongside the methods of data derivation. All tests were performed on uniaxial testing machines. In addition to the acquisition of forces and displacement signals, all experiments were filmed by a CCD camera. Whenever possible, the captured image material was evaluated with a DIC system in order to determine local displacement behaviour. For this reason, all specimens had to be spray-painted with a stochastic grey-scale pattern in preparation of the testing process.

4.3.2.1 Testing procedures and conditions

During the tests constant environmental conditions were ensured. Room temperature was kept at 23 °C whereas humidity was maintained at 50 %. Examples of testing configurations that were used in this work are presented in figure 4.1. Details on the set-up of mechanical tests can be found in figure A.4 in the appendix. Tensile tests were conducted in reference to standard DIN EN ISO 527, while compression tests were performed in reference to standard DIN EN ISO 604 and bending properties were determined in relation to DIN

EN ISO 178. However, the standards were not completely complied with, due to some inherent drawbacks as discussed e.g. in BECKER, 2009.

All quasi-static tensile and shear tests were conducted on a static AllroundLine universal materials testing machine by Zwick Roell AG (Z020) and a static AllroundLine universal materials testing machine by Zwick Roell AG (Z250), using a 20 kN, respectively 100 kN load cell. For the quasi-static testing of compression and three point-bending a static zwickiLine materials testing machine by Zwick Roell AG (Z005) with a 5 kN load cell was applied. Quasi-static tensile, shearing and bending behaviour was evaluated at a displacement rate of 1 mm/min. In case of biaxial loading, quasi-static results were determined at velocities of 6 mm/min. Due to the small scale of compression samples and an effort to minimise friction induced specimen barrelling, compression testing was performed at 0.2 mm/min.

High velocity tests were performed on a high speed servo-hydraulic HTM materials testing machine by Zwick Roell AG (HTM5020). The machine includes a pre-adjustable acceleration unit which ensures that the initial velocity upon loading of the sample complies approximately to the predefined nominal speed. A more detailed description of the set-up can be found in BECKER, 2009. Dynamic testing was done at speeds of 0.1 m/s, 1 m/s and 3 m/s , in order to obtain extended information on strain rate dependent behaviour.

Tensile specimens were mounted into the quasi-static testing machines by a pneumatic clamping system. In case of dynamic testing a manual fixture was applied, in order to avoid high component mass in the acceleration zone. Prior to testing, shear samples were inserted into a special fixture that could be positioned into the tensile set-up, BECKER, 2009.

A puncture test was applied to test the behaviour under equi-biaxial tension. The plate samples where fixed by a circular clamping system and the load was applied at the centre of the sample plates by a piston. The piston was lubricated with Vaseline in order to decrease friction effects.

For compression testing, the guided tension-compression-redirection construction, seen in figure 4.2, was applied. Thereby, axial offset of the machine during compression as well as stability problems were avoided. Special measures were taken to reduce friction. These included application of polytetrafluoroethylene (PTFE) and Vaseline to the friction surfaces. Details on the topic of friction reduction for compression testing can be found in DILLENBERGER, 2011.

Forces were evaluated by load cells positioned in between machine traverse and sample fixture, respectively piston. Displacements during quasi-static testing were either measured by machine traverse movement, a capacitive displacement transducer or a laser sensing unit. The additional displacement measurement systems where positioned between specimen clamps, or between the fixture and a reasonable reference position. Additionally, displacement values were determined optically using grey scale correlation results. In case of dynamic testing, the set-up only permitted the analysis of incremental piston movement for the determination of displacement.

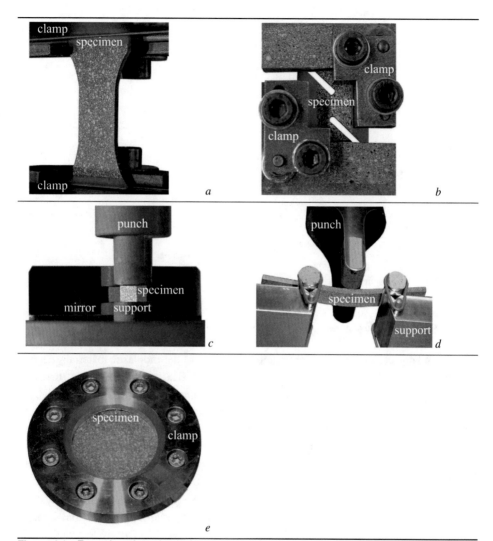

Figure 4.1 Examples of mechanical tests: *(a)* tension, *(b)* shear, *(c)* compression, *(d)* three point bending and *(e)* puncture test.

For each test configuration, the final results were generated from arithmetic average values of five single tests. For biaxial tests, the number of samples was reduced to three, due to a lack of injection moulded parts.

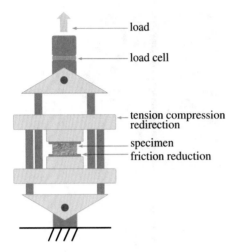

Figure 4.2   Compression test construction.

4.3.2.2 Strain Analysis

For an adequate determination of deformation properties of thermoplastic materials, details on local strains are required, see e.g. JUNGINGER, 2002, BECKER, 2009 and SCHÖPFER, 2011. In this work, local surface strains were determined by digital image correlation (DIC), which was applied in all uniaxial load cases. In comparison to alternative strain measurement techniques, DIC bears the advantage of being contactless. Therefore, it has no physical backlash on the test results. For all quasi-static experiments except compression, the gray scale based 3D DIC hardware and software system ARAMIS 4M by Gesellschaft für optische Messtechnik mbH was used.

The capability of determining 3D displacements by the use of a stereo camera system is particularly important when movement in camera axis, or non-planar, spatial sample surfaces are to be resolved. In this context, a two dimensional (2D) DIC system cannot differentiate between motions in camera-axis and strains transverse to camera axis. Movement of a sample in camera axis would lead to image magnification with usual camera lenses and thus be interpreted as an increase of strain.

In high speed testing of planar specimens, a 2D system was used. A high speed camera, capable of taking up to 125.000 images/s, was utilised for data acquisition. Image rate was adjusted for each testing set-up to achieve a compromise between image and increment size, since larger images reduce the possible recording rate.

Compression testing also required the use of 2D DIC, since samples sizes were too small for the available ARAMIS 4M system to be applicable. In this case, a single monochromatic CCD camera with a resolution of 1280 px × 2024 px, a colour depth of 8 bit and a pixel-size of $8\,\mu m/px$ captured the required images with an average frequency of 1 image/s. To account for the 3D deformation of compression samples, two sides were filmed by the use of a mirror, see figure 4.1 *(c)*. The results from both images were processed in order to determine proper strain values in all three dimensions, as is detailed in DILLENBERGER, 2014. Images acquired by 2D DIC were correlated subsequently using the correlation software Vic2D from Limes GmbH.

Examples of typical grey scale patterns on tension specimens superimposed by longitudinal strain results are shown in figure 4.3. The DIC software uses correlation methods to analyse the distortion of a grey-scale pattern between images of consecutive deformation steps and a predefined, unloaded reference image. For this purpose, the grey-scale pattern is divided into facets, uniquely determined by their respective grey value distribution. These facets are then identified in every image acquired during the loading process. From the deformation of the facet pattern, a pixel based strain measure is calculated. Since the correlation results refer to pixel differences, they have to be calibrated to the actual specimen geometry. Details on the application of grey-scale pattern based DIC in the context of mechanical experiments can be found in e.g BECKER, 2009 and PALANIVELU, S. et al., 2009.

The final result of the correlation is a strain field over the whole area of interest on the specimen surface, calculated from the lateral and longitudinal deformation of the grey-scale pattern, see figure 4.3. In case of a 3D analysis, additional information on facet displacement in thickness direction, respectively camera axis orientation, is available. Strains are calculated logarithmically as presented in equation (3.18).

A scalar reference strain measure needs to be derived from the lateral and longitudinal 2D strain fields in order to identify a distinct stress-strain relation for the evaluation of material parameters. This was achieved by the utilisation of an averaging area as proposed in BECKER, 2009. In order to gain comparable data over all tests, a fixed rectangular area was defined for each test set-up. Since the principal interest lies in material performance at the section of maximum deformation and failure, averaging areas were chosen accordingly. A detailed visualisation of the applied strain averaging areas is given in figure A.4 in the appendix.

In case of tensile specimens, strain averaging areas where selected differently for PP and PPGF30 materials. Brittle PPGF30 specimens show a highly non-uniform strain distribution with local strain peaks due to embedded fibres, see 4.3 *(b)*. Thus, a large averaging region of $12\,mm \times 12\,mm$ over the entire parallel gauge length was chosen. This allowed the derivation of more robust average values for all test specimens that are less influenced by singular strain peaks.

In case of non-reinforced PP, specimens show a distinct strain localisation followed by a more uniform strain distribution, see 4.3 *(a)*. In order to adequately take into account strain localisation, a $12\,mm \times 2\,mm$ averaging area in the middle of the specimen was chosen. For shear tests, an evaluation area of $2\,mm \times 6\,mm$ over the entire shearing zone of the samples was applied. In the case of compression testing, an averaging zone of 2 mm × 1.2 mm on each side of the sample was utilised, thus permitting a measure over almost the complete sample surface, while neglecting boundary effects.

For a validation of specimen deformation in a finite element simulation context, the averaging areas in the virtual experiments would have to be chosen accordingly to allow for an adequate comparison with test data.

The scalar strain values derived by the area averaging process can be used for the preparation of strain curves required for material parametrisation. Experimental strain information in this work is always taken from local DIC analysis, and refers to logarithmic true strain values if not otherwise indicated.

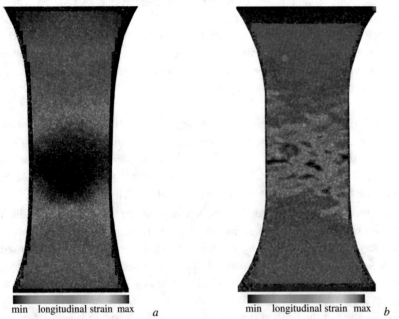

min   longitudinal strain max   *a*   min   longitudinal strain max   *b*

Figure 4.3 Examples of grey scale patterns overlayed with DIC results for longitudinal Hencky's strain: *(a)* non-reinforced PP, *(b)* SFRTP PPGF30.

4.3.2.3 Stress Analysis

Stress values were determined by relating data from the machine-integrated force sensors to sample cross sections. Due to the testing set-up, stress evaluation was only possible for tension, shear and compression. Calculation of engineering stress values was performed by application of the initial cross-section A_0 as stated in equation (3.25). However, since information on local strain deformation is retrievable, assumptions can be made considering the actual specimen cross section. Stress values related to the current cross section are comparable to those that are needed in an FEA simulation since they give information on locally effective stresses. They are thus needed for a proper parametrisation of material behaviour. In case the principal strain system coincides with the axis of load application, the expression for true stress given in equation (3.26) can be adapted to

$$\sigma = \frac{F}{A_0 \exp(\varepsilon_x + \varepsilon_z)}. \tag{4.1}$$

Herein, the deformation of the actual sample cross-section is described by the lateral strain ε_x and the through-thickness strain ε_z.

Expression (4.1) requires information on the actual cross section A, respectively the actual thickness related Poisson's ratio. Commonly, it is assumed that in-plane lateral strains are equal to the strains in through-thickness direction of the sample by $\varepsilon_z \approx \varepsilon_x$. This results in

$$\sigma^{2D} = \frac{F}{A_0 \exp(2\varepsilon_x)}, \tag{4.2}$$

which is henceforth termed 2D stress measure or transverse strain equality assumption.

Expression (4.2) can be further simplified by assuming an approximately isochoric sample deformation with $v_{xy} = 0.5$. This leads to the isochoric (isoc) stress expression

$$\sigma^{\text{isoc}} = \frac{F}{A_0 \exp(-\varepsilon_y)}, \tag{4.3}$$

that is only depending on longitudinal strain.

Equation (3.21), that is applied in expressions (4.1) and (4.3) in order to determine the change of lengths in the cross section, is only valid in the principal strain system, compare section 3.1. Thus, for anisotropic materials or off-axis tests with orthotropic materials these expressions may not be applied. In this case, the true strain tensor needs to be transformed into its principal axis system determined by its eigenvectors. In the principal axis system the engineering strains are computed and then transformed back to the original coordinate system. From the resulting values the change of the cross section is determined.

In theory the assumption of expression (4.2) may hold for (transversely) isotropic materials, but is questionable for orthotropic materials such as the SFRTP composites in concern. This is due to the fact that such an approximation would only be valid for materials where the amount of fibres in lateral and thickness direction is equal, which is generally not true for injection moulded composites except in special cases, cf. section 2. SCHÖPFER, 2011 and VOGLER, 2014 suggest the relation $\varepsilon_z \approx 1.7\varepsilon_x$ to account for a higher thickness straining potential, respectively a lower fibre orientation in thickness direction. It is clear that such an assumption cannot hold for every SFRTP composite, since the fibre orientation may differ, depending on material and mould. RÖHRIG et al., 2017 propose a correlated two-sided strain evaluation for the simultaneous characterisation of transverse and thickness strains. Nevertheless, the suggested method requires a complex setup of two cameras on the front and two cameras on the rear side of the specimen.

Therefore, in the context of the present study, a modified strain measurement method was established that allows for the evaluation of thickness strains under the assumption that the deformation of tensile samples in thickness direction occurs symmetrically. Since the applied 3D DIC method only enables analysing the sample from one side, the thickness strain measurement was obtained by a detailed evaluation of the overall sample motion.

In figure 4.4 *(b)*, the twisting and skewing of a loaded sample by the offset of clamps of the machine is shown. The amount of these effects is scaled up for visualisation purposes. As visible in the figure, the overall displacement of the surface evaluation area on a loaded tensile sample is a superposition of rigid body translation, rotation and straining. The one sided DIC method can directly calculate strains that occur laterally to the axis of the image acquisition camera. However, since the one sided DIC analysis cannot distinguish between translation, respectively rotation, and actual strains in the camera axis z, it will only give information of the displacement u_z of measurement points in this direction.

By adding multiple rigid body markers to the measurement set-up, overall movement of the sample during loading can be analysed in detail. The markers are applied to the clamps that are expected to experience negligible straining in respect to the sample. This set-up gives the possibility of calculating a rigid body reference plane, spanned between the rigid body markers, for each evaluation step. By comparing the movement of the reference plane $u_{z,0}$ to the displacement of points in the evaluation area u_z, assuming symmetrical displacement in thickness direction and relating the result to the initial sample thickness s_0, the actual logarithmic thickness strain ε_z can be extracted by applying equation (3.21) and setting

$$\varepsilon_z = \ln\left(1 + \varepsilon_z^{\text{eng}}\right) = \ln\left(1 + 2\frac{u_z - u_{z,0}}{s_0}\right). \tag{4.4}$$

It has been discussed in section 3.1 that this computation of true strains from engineering strains is only valid on an axis of principal strain. Therefore, equation (4.4) applies if the material is isotropic or orthotropic and it is assumed that the through thickness direction coincides with a principal axis of the material coordinate system.

The overall measurement set-up for the evaluation of tensile sample deformation is shown in figure 4.4. It allows for an improved evaluation of the actual cross section by means of a standard 3D DIC set-up.

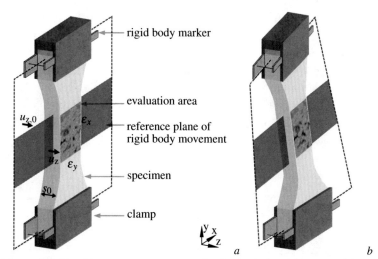

Figure 4.4 Tension measurement set-up for the analysis of thickness deformation: *(a)* unloaded sample and *(b)* loaded sample exposed to straining, additionally being skewed and twisted by an offset of the clamps.

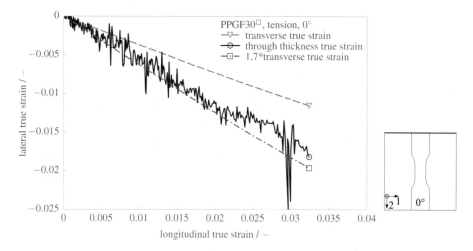

Figure 4.5 Thickness strain evaluation.

Figure 4.5 shows the result of the thickness strain measurement compared to transverse strain values and the approximation given in SCHÖPFER, 2011. The thickness strain shows a large amount of scatter in comparison to transverse strain data. On the one hand this is due to the sensitivity of the measurement method, and on the other hand this is the result of the evaluation method which involves trigonometric calculations with small strain values. If displacement differences are small, the trigonometric equations may lead to non-physical high values. This could be overcome by an adequate filtering method. It can be seen that at the beginning of the deformation process, for longitudinal strain values < 0.005, thickness and transverse strain are in a very similar range. Due to scatter of results it is not clear if thickness strain is higher or lower than transverse strain in this region. With ongoing deformation the thickness strain increases more than the transverse strain. Furthermore, figure 4.5 shows that the assumption made by SCHÖPFER, 2011 gives a good approximation of the thickness strain at larger deformations for the tensile samples extracted from the injection moulded SFRTP plates.

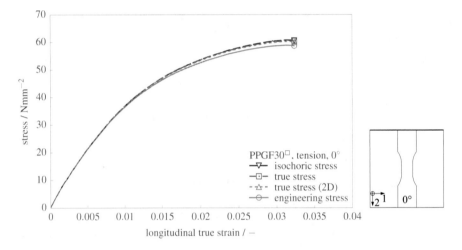

Figure 4.6 Comparison of different stress evaluation techniques for PPGF30.

The scalar strain values derived by the averaging process and the calculated stress values can be combined to stress-strain curves. However, since the total strain to fracture is of small magnitude, the actual cross section reduction of the specimen has only limited influence on the total amount of stress. This becomes evident upon inspection of figure 4.6. The consideration of a reduced cross section only leads to slightly increased values in respect to engineering stress. Moreover, while values gained from the measurement of actual thickness strain lay in between the transverse strain equality and the isochoric deformation assumption, the differences can be considered insignificant. With regard to experimental effort, it can be seen that the novel measurement method can be neglected. Even the transverse strain equality assumption does not give noticeably different results to the isochoric one. From the viewpoint of measuring an adequately exact stress, it would hence be sufficient to solely evaluate local longitudinal strains. Furthermore, evaluation of engineering stress values may be acceptable for some applications. These assumptions hold for materials with comparably small total deformation capabilities as the SFRTP evaluated in this work.

The proposed thickness measurement method is of much larger interest for materials with high straining capabilities that show significant necking, such as non- or minimally reinforced polymers. This is emphasised by figure 4.7, which shows the previously introduced techniques for the PP material. It can be noted that the stresses evaluated by use of the actual cross section strongly deviate from those gained by the 2D, respectively transverse strain assumption. The actual cross section measurement shows that in the case of PP, the isochoric assumption depicts the material behaviour at high deformation more satisfactorily. This is especially relevant in FEA simulations when material models are applied that take into account actual plastic Poisson's ratio. An improper designation of thickness in respect to lateral strains could lead to erroneous results in this case. The high scatter at the end of the deformation process in figure 4.7 is a consequence of the highly distorted grey scale pattern due to the large straining of the PP material. DIC fails due to too distorted facets, thus leading to the perceived scatter of strain values. Consequently, the scatter is higher if multiple strain measures are taken into account for stress evaluation. Such is the case for true stress and true stress 2D values. Isochoric stress is only influenced by one strain variable, and therefore shows lower scatter. Since no strain value is used for the derivation of engineering stress, no scatter influences this stress value.

The choice of an adequate stress evaluation method depends on the capabilities of the material model that needs to be parametrised for subsequent simulation steps. For material models that allow the input of true stress values and have the capability of reproducing non-isochoric plastic deformation, stress evaluation during experiments has to be done accordingly. When this is not the case, an evaluation of isochoric stress values is sufficient.

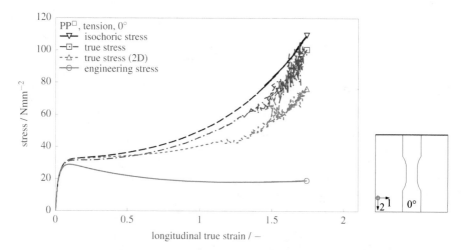

Figure 4.7 Comparison of different stress evaluation techniques for PP. Scatter at large deformations occurs due to failing DIC facet recognition.

4.3.2.4 Material Parametrisation

In the context of material modelling, a derivation of elastic and plastic material properties from the overall material behaviour is necessary. Thermoplastic materials mostly do not show a pronounced yield point, respectively switch-over point from elastic to plastic behaviour, as various metallic materials do. This has been elaborated in chapter 1. Therefore, the term yield for thermoplastic materials considered in this work is used for the indication of the bounds of linear behaviour. A yield point will herein describe the switch from linear to non-linear behaviour of the material. Considering material modelling, linear behaviour is assumed to be purely elastic while the beginning of non-linearity is assumed to introduce plasticity. This assumption does not hold in a strict physical sense but can adequately represent material behaviour in a material modelling context. It is widely used for the definition of material models for thermoplastics in crash applications, see e.g. BECKER, 2009, GRUBER et al., 2011, SCHÖPFER, 2011, VOGLER, 2014 and others. The assumption has also been applied in quasi-static modelling approaches, see e.g. KAISER, 2013. Following the concept of the aforementioned researchers, non-linear viscoelasticity is neglected in the context of this work, even though it can be of perceivable influence in thermoplastic components. The procedures described in the following are visualised in figure 4.8.

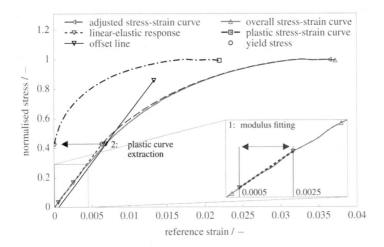

Figure 4.8   Parametrisation of stress-strain curves.

Determination of Young's modulus

The linear elastic region in the stress-strain curves is described by Hooke's law. Following the approach given in DIN EN ISO 527, a straight line was fitted to the stress-strain response in the small strain region of 0.05 % to 0.25 % by a least squares method. The slope of the fitted line determines the Young's modulus of the material. Poisson's ratio is evaluated accordingly.

Determination of yield

During literature review it was noticed that often a precise definition of yield point evaluation is missing, presenting an obstacle for the reproducibility of research results. This is of even more significance upon the comparison of experimental data with virtual results determined from modelling equations, where a clear account on the prediction quality of a specific model can only be made, if experiments are evaluated in a non-random and reproducible manner. For the evaluation of metal components without a pronounced yield point, the bounds of linear behaviour are defined by an offset yield strength, as described e.g. in BARGEL et al., 2008. It is found by the intersection of a line defined by Hooke's law that is offset on the strain axis by an arbitrary value. Thereby, the problem of finding an explicit distinction point in the continuous transition from linearity to non-linearity is avoided and reproducibility of results is ensured. This method was for example applied by BOSSELER, 2017 to derive yield values for long glass fibre reinforced PP.

A similar concept was followed in this work. The offset yield strength was herein evaluated for the composite materials at an offset strain value ε_{rp}. For a reproducibility of results, the offset needs to be set to a value that is used for the evaluation of all experiments. Since a literature review gave no conclusive value for such an offset strain for SFRTP materials, the value was set to $\varepsilon_{rp} = 0.0003$ in this work.

The utilisation of an offset yield strength will lead to a strain offset at the yield point between the end of the straight line defining linear-elastic behaviour and the beginning of the non-linear curve determining plastic behaviour. This non-physical gap is solely introduced by the evaluation technique. In order to achieve a continuous transition and to overcome the gap, an adjustment of the non-linear curve is suggested in this work. Thus, the Young's modulus needs not be altered. A simple continuous scaling of the strain-axis of the non-linear curve is proposed by

$$\varepsilon^{\text{adjusted}} = \varepsilon f(\varepsilon). \tag{4.5}$$

Three constraints have to be met by the scaling function $f(\varepsilon)$:

- Firstly, the scaling at yield has to lead to a strain equality at the end point of the linear curve and the starting point of the non-linear curve. The former is defined by Hooke's law and results in σ_Y/E. The latter is depends on the predefined offset strain value ε_{rp} by $\varepsilon_0 = \sigma_Y/E + \varepsilon_{rp}$. This leads to an initial strain scaling factor of $f_0 = \frac{\sigma_Y/E}{\varepsilon_0}$.

- Secondly, values of the fracture strain ε_F, respectively the end of the curve must not be changed. Thus, a final strain scaling factor of $f_F = 1$ will leave fracture strain of the non-linear curve unaffected.

- Lastly, the scaling factor must diminish from f_0 to f_F over the complete strain range of the non-linear curve.

Applying the previous constraints, the expression

$$f(\varepsilon) = \frac{f_F - f_0}{\varepsilon_F - \varepsilon_0}(\varepsilon - \varepsilon_0) + f_0 \qquad (4.6)$$

is proposed for the scaling factor as a function of ε. In this linear formulation, the scaling factor will diminish to the value f_F at ε_F. An example of the proposed continuous scaling can be seen in figure 4.8 in the adjusted stress-strain curve.

Determination of plastic strains
Once a continuous stress-strain response has been recovered, plastic behaviour can be evaluated in more detail. In this work, elasto-plastic decomposition is assumed, compare chapter 3.4. Hence, including the previous definition of yield, total strain is expected to be a sum of linear elastic and non-linear plastic strains, as shown in expression (3.49).

Therefore, uniaxial plastic stress-strain responses, as given in figure 4.8, can be attained from total stress-strain curves by relating stresses succeeding the yielding point to plastic strains. Plastic strains are determined by expression (3.51) which is the result of combining Hooke's law and equation (3.49).

Due to strain equivalence as defined in chapter 3.6 this expression holds for the undamaged as well as the damaged configuration. In the case of damage, stress as well as Young's modulus become dependent on the damage variable d.

4.3.2.5 Consideration of Damage

In an experimental context, the damage variable can be quantified by measuring the stiffness softening as stated in equation (3.61) in relation to plastic strain. Samples are cyclically loaded to different values of total strain and the modulus of the damaged configuration $\tilde{E}$ is analysed upon unloading. The overall modulus reduction in comparison to the initial Young's modulus of purely elastic loading should be related to the plastic strain of each step in order to determine the damage variable. However, the calculation of a plastic strain value by use of (3.62) is not possible in a straightforward manner, since the damaged modulus $\tilde{E}$ used herein depends on the damage variable d that then again depends on plastic strain ε^{pl}. This leads to the equation

$$\varepsilon^{\text{pl}} = \varepsilon - \frac{\tilde{\sigma}}{\tilde{E}} \quad = \varepsilon - \frac{\tilde{\sigma}}{E(1 - d(\varepsilon^{\text{pl}}))}. \qquad (4.7)$$

A numerical optimisation scheme can be applied to expression (4.7) in order to find a plastic strain value solving the equation.

4.3.2.6 Evaluation of Strain Rate Dependency

As seen in figure 4.3 strain localisation effects are distinct in non-reinforced polymers while being scattered in SFRTPs. Thermoplastic materials show no constant rate of strain in the course of deformation due to these localisation effects. This is especially the case at high rates of displacement given the small amount of time the material is provided in order to develop plastic flow. However, in the context of material modelling stress-strain data at constant strain rates is required. BECKER, 2009 introduced a procedure of determining these curves. This process was applied on the experimental results of this work and shall be summarised in short in the following. For further details refer to BECKER, 2009.

At first, stress-strain curves of equal tests performed at different displacement velocities are plotted in a 3D diagram depicting stress, strain and strain rate. In the next step, a model equation capable of mapping strains and strain rates to stress values has to be selected. In BECKER, 2009 multiple models are compared out of which the stress expressions by Johnson-Cook σ^{JC}, G'Sell-Jonas σ^{GJ} and Cowper-Symonds σ^{CS} are given as

$$\sigma^{JC}(\varepsilon_n, \dot{\varepsilon}) = \sigma_0(\varepsilon_n) \left(1 + \zeta^{JC}(\varepsilon_n) \ln \left(\frac{\dot{\varepsilon}}{\dot{\varepsilon}_0(\varepsilon_n)} \right) \right), \tag{4.8}$$

$$\sigma^{GJ}(\varepsilon_n, \dot{\varepsilon}) = \sigma_0(\varepsilon_n) \left(\frac{\dot{\varepsilon}}{\dot{\varepsilon}_0(\varepsilon_n)} \right)^{\zeta^{GJ}(\varepsilon_n)}, \tag{4.9}$$

and

$$\sigma^{CS}(\varepsilon_n, \dot{\varepsilon}) = \sigma_0(\varepsilon_n) \left(0.5 + 0.5 \left(\frac{\dot{\varepsilon}}{\dot{\varepsilon}_0(\varepsilon_n)} \right)^{\frac{1}{\zeta^{CS}(\varepsilon_n)}} \right). \tag{4.10}$$

Equations (4.8), (4.9) and (4.10) compute stresses at a significant strain rate $\dot{\varepsilon}$ by scaling stress values σ_0 at a predefined reference strain rate $\dot{\varepsilon}_0$. The values ζ^{JC}, ζ^{GJ} and ζ^{CS} represent fitting parameters to adjust the amount of stress scaling. To achieve best fitting values over the entire strain rate range, the value σ_0 can also be interpreted as fitting parameter. The equations are given in terms of strain increments ε_n since BECKER, 2009 suggests the fitting of measurement data at multiple points of constant strain. In order to achieve this, all measured stress-strain-strain rate curves have to be interpolated at equal strain values. After fitting is completed for every strain point, a set of multiple modelling parameters ζ is available. With these results a curve can be defined at each previously defined point of constant strain which fits stresses and strain rates. The entirety of these curves forms a 3D surface over all measured data. The stress-strain responses at constant strain rates can then be extracted by evaluating the fitted 3D surface at the desired rates. A example of the previously stated process is visualised in figure 4.9 for PP.

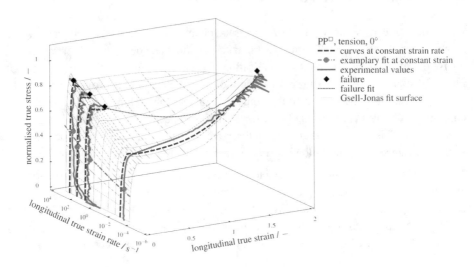

Figure 4.9 Extraction of curves at constant strain rate shown for PP. Data was normalised to the maximum stress value of the quasi-static curve.

For thermoplastic materials the application of the G'Sell-Jonas model is suggested by BECKER, 2009. SCHOSSIG, 2010 tested various short fibre reinforced PP composites at multiple strain rates and also proposed the G'Sell-Jonas model. Accordingly, throughout the present work the G'Sell-Jonas approach is utilised to take into account the strain rate dependency.

Figure 4.9 includes the plot of a fitted line for the depiction of failure strains. The determination of failure strains has to be done separately from the aforementioned construction of a 3D surface from the fitting at constant strains, since failure strains may not be constant in relation to strain rate. The individually fitted line over all failure strains, determined by one of the expressions (4.8) - (4.10), represents an upper bound for the extracted curves at constant strain rates. These are cut off once they reach a value depicted by the failure curve. Thereby, adequate failure values for modelled stress-strain curves can be extracted.

4.3.3 Micro-Structural Analysis

In order to evaluate anisotropic material behaviour of SFRTPs, an analysis of intrinsic microstructural properties is required. As stated in chapter 2.4, the fibre distribution strongly influences the mechanical behaviour of SFRTPs. The following sections summarise the steps taken in order to evaluate the micro-structure of SFRTP components for the purpose of generating reliable information on FODs and FLDs.

4.3.3.1 Testing Procedure

In this work, µCT methods were used to analyse micro-structural aspects of SFRTP components. The samples that were extracted from SFRTP parts were scanned in a SkyScan 1072-100 µCT apparatus. In order to adequately resolve the inclusion fibres, the scan images were generated with the maximum resolution of 1.8 µm/ voxel. Figure 4.10 shows a schematic of the µCT procedure. An actual specimen and an example for the images retrieved from the scanning process can be found in figure 4.11. Samples were exposed to x-rays of low intensity, while being rotated around their vertical axis. X-ray absorption depends on the density of constituents in the sample, hence x-ray magnitude reaching the detector varies with sample orientation. An 180° analysis of a sample was performed at different positions on the vertical axis. This resulted in a voxel based information on the density distribution over sample height and thickness. Slice images in respect to the sample height in z-axis were extracted form this 3D density representation.

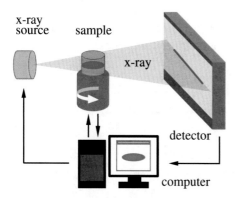

Figure 4.10   Process schematic of µCT analysis.

An overview of different fibre evaluation methods and the respective advantages or drawbacks was given in section 2.5. In this work, the iterative, model-based Monte-Carlo analysis percolation algorithm, presented by GLÖCKNER et al., 2016, was used for the recognition of the fibres from µCT images. It has the advantage of being an automatic detection method that extracts information on the distribution of single fibres over the entire sample volume from the µCT images.

The automatic analysis procedure requires an input of constituent fractions and sample material densities. In the case of PPGF30, the fibre weight fraction was measured by a thermo-gravimetric analysis, in which the thermoplastic sample was incinerated and the weight of remaining fibres was quantified. The resulting weight fraction of 30.68 % closely matched the 30 % fibre weight fraction used in the compounding process.

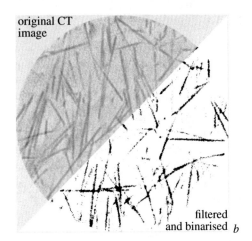

a

Figure 4.11 μCT analysis: *(a)* specimen and *(b)* slice image.

The applied recognition algorithm consists of four process steps, which are briefly summarised in the following. Details on the algorithm and its validation are given in GLÖCKNER et al., 2016.

- μCT images are prepared by applying filtering and binarisation methods, see figure 4.11 *(b)*. Binarisation is done with respect to a fibre volume fraction that needs to be input to the algorithm prior to the analysis.

- Multiple iterative recognition processes are parallelly executed. Each of the algorithmic instances analyses spherical integrals at random positions to identify fibre centres. From the found fibre centres, orientation and length evaluation procedures are initiated. The stochastic nature of the algorithm, as well as process parallelisation, allows multiple detections of the same fibres at different starting points. This leads to an improved recognition quality of single fibres.

- Data from parallel calculations is merged after each recognition process. Duplicate fibres are detected and voxels of reasonably well detected fibres are erased from the input data. Then, the fibre detection algorithm is restarted.

- Once a predefined detection accuracy is reached, all recognition processes are terminated and the geometrical fibre properties are computed.

The detection algorithm assumes that fibres are of constant diameter and cylindrical shape with marginal curvature. The algorithm makes it possible to assign voxels of μCT images to individual fibres in spite of inherent image noise. Upon completion, the recognition software gives out the spatial point of origin, the orientation vector, the length and the radius of every detected fibre. The evaluated information enables the reconstruction of local fibre distributions from μCT images to the degree of detection accuracy.

An exemplary, reconstructed micro-structure is given in figure 4.14 *(a)*. In this work, further evaluation of micro-structures is always performed by using the reconstructed data. Accordingly, the presented recognition method can be seen as a preprocessing of measured data.

Besides using μCT methods and the presented recognition procedures for the quantitative analysis of micro-structures, SEM images where used for the qualitative evaluation of fracture surfaces of specific tensile specimens.

4.3.3.2 Data Evaluation

Orientation tensors are used for the specification of FODs in this work. Details on orientation tensors have been given in section 2.5. For the determination of tensor values from experimental fibre data, evaluated by the previously described μCT analysis method, expression (2.6) is applied. Following the suggestions of GLÖCKNER et al., 2016 for the experimental derivation of orientation tensors, the fibre volume is used as weighting factor for the single fibres in equation (2.6), as has been discussed in section 2.5. The average fibre aspect ratio in a specific volume section was evaluated by equation (2.9).

In order to graphically visualise properties of the FOD, respectively the second order orientation tensor **A** introduced in section 2.5 equation (2.6), most commonly diagonal tensor components in a reference system are compared in a 2D plot. More information on this visualisation technique and examples can be found in e.g. MLEKUSCH, 1999, KAISER, 2013 and MÖNNICH, 2015. In MÜLLER et al., 2015 a detailed presentation on the visualisation of different aspects of the second order orientation tensor can be found and it was shown that upon the visualisation of solely diagonal components it must be ensured that the principal axes of the FODs are well defined. It is problematic to use such a method in an experimental context when principal axis of FODs and direction of flow, respectively the coordinate system of samples under consideration, are not parallel. If this is the case, evaluation results of diagonal tensor components in main flow direction are misleading and hardly interpretable, since non-zero off-diagonal components are neglected. Even when the direction of flow and the principal axis system concerning the global orientation tensor of a sample are co-linear, it is insufficient to simply visualise the diagonal components when through-thickness distributions of orientation tensors are to be analysed. The principal axis system may change for every point of considered in the through-thickness evaluation. This may result in non-zero off-diagonal components of the local orientation tensor, even if the global orientation tensor considering the entire specimen only has non-zero values on the main diagonal.

An alternative analysis of the diagonal components strictly in each principal axis system, would give no information on their relation to the main flow direction. Nevertheless, this is of interest in an engineering context.

Due to these drawbacks, an adjusted visualisation technique is introduced in this work. In this context, the ellipsoidal representation in figure 2.10 is taken into account. In section 3.2 the terms eigenvector $\vec{q}$ and eigenvalue λ of a tensor have been explained. The main direction of preferred fibre orientation, specified by the first eigenvector $\vec{q}_1$, in respect to a reference coordinate system can be specified by its angles ϕ and θ. Reference axis $\vec{e}_2$ is determined by the main direction of polymer flow during injection moulding in a component. Axis $\vec{e}_1$ of the reference system is orientated transversely to axis 2 and parallelly to the component surface. The remaining axis $\vec{e}_3$ is oriented in through-thickness direction of the component, normal to the plane spanned by axes 1 and 2. In the first step ϕ and θ are normalised by a quadratic sine function. Application of the quadratic function leads to a simpler and more practical, distinct visualisation, since negative angle values are avoided and more weight is given to large deviations from the reference axis in contrast to small ones. Overall, the main fibre orientation in respect to the reference system is described by the planar orientation measure

$$\sin^2 \phi \begin{cases} = 1 & \rightarrow \text{parallel to axis 2} \\ = 0 & \rightarrow \text{transverse to axis 2,} \end{cases} \tag{4.11}$$

and the spatial orientation measure

$$\sin^2 \theta \begin{cases} = 1 & \rightarrow \text{parallel to axis 3} \\ = 0 & \rightarrow \text{transverse to axis 3.} \end{cases} \tag{4.12}$$

The degree of anisotropy, respectively probability of fibres oriented in the main direction is defined by the roundness of the ellipsoid in figure 2.10. The roundness of the ellipsoid is determined by its eigenvalues λ. The ellipsoid will expand to a unit sphere in case of a random FOD. In a highly orientated FOD it will collapse to a needle-like form. In a practical context, a single scalar value for the definition of an anisotropy degree is preferable. One option of describing the tensor $\mathbf{A}$ by scalar values is the computation of its corresponding tensor invariants. These are defined similarly to the invariants of the stress tensor presented in section 3.2. In the principal axis system of the second order orientation tensor $\mathbf{A}$ the invariants can be expressed as

$$\begin{aligned} I_1(\mathbf{A}) &= \lambda_1 + \lambda_2 + \lambda_3, \\ I_2(\mathbf{A}) &= \lambda_1\lambda_2 + \lambda_2\lambda_3 + \lambda_1\lambda_3 \quad \text{and} \\ I_3(\mathbf{A}) &= \lambda_1\lambda_2\lambda_3, \end{aligned} \tag{4.13}$$

compare ALTENBACH, 2015.

Hence, the tensor invariants are related to the form of the ellipsoid, since they are proportional to its specific geometric characteristics, as displayed in figure 4.13. This leads

to the conclusion that a tensor invariant can adequately describe the anisotropy degree of
A. Since the orientation tensor is normalised, as seen in section 2.5, the first invariant will
always result in the value 1 which makes it inadequate for this purpose. The choice of an
alternative invariant for the description of anisotropy is arbitrary. In this work, the second
invariant was chosen, since it corresponds to the generalised Herman's orientation given as
anisotropy degree measure in ADVANI et al., 1987. The invariant value is scaled to give
values between zero and unity, resulting in the expression

$$\alpha = \frac{3}{2}I_2(\mathbf{A}) - \frac{1}{2} \begin{cases} = 1 & \rightarrow \text{completely aligned fibres} \\ = 0.5 & \rightarrow \text{planar isotropic} \\ = 0 & \rightarrow \text{randomly distributed fibres.} \end{cases} \tag{4.14}$$

for the anisotropy degree α.

Figure 4.12 shows resulting anisotropy degrees for different eigenvalue constellations.
Depending on the deviation from planar distributions, given by the value $\lambda_3 \geq 0$, anisotropy
degrees are displayed for different principal eigenvalues λ_1. Figure 4.12 takes into account
that the sum of all eigenvalues is always 1 by definition of the orientation tensor. Therefore,
the respective values for λ_2 are not explicitly given.

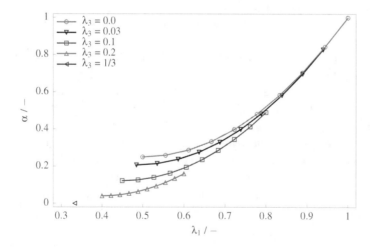

Figure 4.12 Exemplary values for the anisotropy degree α in relation to different eigenvalues.

In summary, the final values for the planar orientation $\sin^2 \phi$, the spatial orientation
$\sin^2 \theta$ and the anisotropy degree α will be used to visualise FODs, respectively second
order orientation tensors throughout this work. The visualisation of these three aspects
leads to a more comprehensible description of FODs involved and overcomes the previously

described inadmissibility of the diagonal component visualisation. A comparison of both visualisation types is given in section 4.4.1.

$$I_1 \propto \lambda_2 + \lambda_2 + \lambda_3$$

$$I_2 \propto A_1 + A_2 + A_3$$

$$I_3 \propto V$$

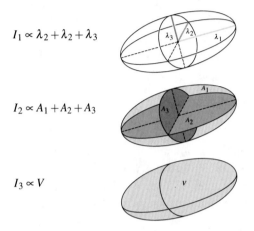

Figure 4.13 Geometrical interpretation of invariants of symmetrical second order tensors. The invariants are proportional to respective geometrical characteristics of the tensor ellipsoid.

For a detailed inspection of fibre distributions over specimen thickness in the z-axis, samples which were scanned and evaluated by μCT were virtually sliced into sections of equal height. Herein, fibres are assigned to a respective slice by their midpoint. Each of these virtual slices is evaluated separately concerning the orientation tensor **A** and the average aspect ratio φ. Thus, information of local FODs and FLDs is presented in respect to sample thickness. An example of a virtually subdivided specimen is given in figure 4.14 *(b)*. While only five slices are presented in the figure for a clearer visualisation, actually twenty slices were used for the evaluation of micro-structures in this work.

Evaluation of fracture surface micrographs retrieved by SEM was done visually on a purely qualitative basis.

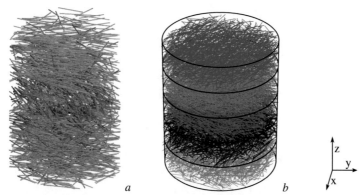

Figure 4.14 Fibre recognition: *(a)* reconstructed micro-structure and *(b)* subdivided fibre information in specimen thickness direction. Through-thickness direction of samples is designated by the z-axis.

4.4 Experimental Results

This chapter contains the experimental results with focus on specific material characteristics of short fibre reinforced thermoplastics. The target of the experiments was not the evaluation of absolute stress-strain values, but rather a relation of anisotropic characteristics to the material response. The data was used to develop the final material model. This material model shall be generally applicable to the class of short fibre reinforced materials and not only specifically to the PPGF30 material inspected. Most of the experimental data was hence normalised with respect to a specific reference state, allowing the emphasis of relative differences. For each material the quasi-static response of respective tensile samples extracted from the plate □ with the dimensions 80 mm × 80 mm × 2.5 mm at position P2 and loaded in a direction of 0° is defined as reference. The true stress-strain response from this reference configuration is defined as reference curve. It is given in figure 4.16. In case of depicted normalised stress values, unless otherwise stated, the maximum true stress value of the reference curve is used for normalisation. Regarding stiffness evaluations, the modulus of the reference curve is applied as normalisation value. This approach clarifies mechanical behaviour differing from the reference state in case of a variation of the loading direction or the intrinsic fibre orientation. In case a different normalisation method was applied, this is noted explicitly in the axis labels.

Wherever error bars are shown, they represent the double standard deviation measured in the experiments.

4.4.1 Anisotropic Stress-Strain Behaviour

In order to determine the influence of the matrix on total anisotropy of the PPGF30 composite, the matrix material PP was tested in different extraction orientations. The testing

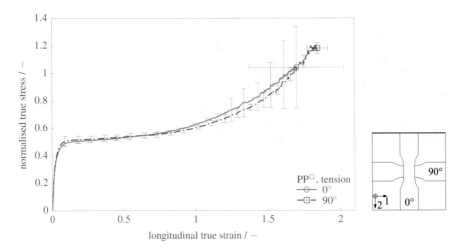

Figure 4.15   Extraction orientation influence on stress-strain behaviour of non-reinforced PP$^\square$.

was done with samples extracted from the 80 mm × 80 mm × 2.5 mm part ($\square$). Results for
the PP$^\square$ part are shown in figure 4.15.

It can be seen that variation of extraction direction has only minor effects on the overall
stress-strain behaviour of the matrix material. Deviation of stress values between the two
directions lay in the range of scatter of the single experiments. For this reason, the matrix
influence on overall anisotropy is considered to be negligible and PP is approximated to
be isotropic. The PP$^\square$ part shows ductile behaviour with a significantly high amount of
deformation. As discussed in section 4.3.2.3, the scatter of stress values at large deformations occurs due to failing DIC facet recognition. This is the case e.g. when DIC facets are
overly distorted. A logarithmic fracture strain of nearly 2 is reached before fracture occurs.
Concerning these values, a noticeable difference can be perceived in the scatter of results.
90° experiments show more reproducible results than those from 0° samples.

Figure 4.16 shows the results of the tensile tests performed on the SFRTP plate part
PPGF30$^\square$. An additional orientation of 45° was checked next to the lateral and longitudinal
samples. In comparison to the behaviour of the PP$^\square$ part, the composite material shows that
the reinforcement with short glass fibres leads to a significant increase of material stiffness
and strength.

A dependence of the results on the extraction orientation is determinable. This mechanical anisotropy is caused by the fibre distribution in the injection moulded plates. It is
noticeable that the stress-strain curves of 90° and 45° are in a similar range. Longitudinal
behaviour at 0° shows higher stiffness and strength values, due to the fact that more fibres
are oriented in the load path. Overall, 90° loading gives the weakest stiffness and strength
response. The 45° curve shows similar stiffness to the 90° case, but higher stress values at

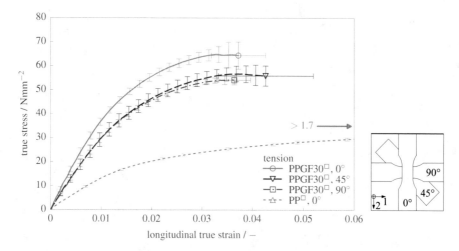

Figure 4.16 Extraction orientation influence on stress-strain behaviour of PPGF30$^\square$.

non-linear plastic deformations. While the PP$^\square$ part shows ductile behaviour, the reinforced composite has brittle characteristics. Fracture strain is strongly reduced, being only around 2 % of the original failure strain of PP$^\square$. 45° samples show higher average fracture strain values and 0° samples slightly lower ones than the 90° samples.

Overall, the anisotropic mechanical characteristics of the samples extracted from the plate are not strongly pronounced and the differences between the maximum stress of 0° and 90° samples lay in the range of 16 %. The reason for this behaviour is found in the fibre distribution in the plate. Figure 4.17 depicts the FODs in different positions over the direction of flow during injection moulding. For the FOD visualisation the methodology described in section 4.3.3.2 is used. The fibre distribution varies over the 20 through-thickness evaluation points.

As discussed in chapter 2.4, the fibre orientation of injection moulded parts depends on the material properties, e.g. pressure dependency, temperature dependency, density and viscosity, as well as injection moulding processing parameters and mould geometry.

The FOD of the plate shows a skin core morphology typically found in injection moulded shell like parts. The upper part of figure 4.17 displays the planar orientation of fibres and shows that the majority of fibres are oriented in main flow direction, respectively 2-axis, in the upper and lower boundary layer of the plate. In the core layer fibres are mostly oriented transversely to the direction of flow. In between layers with lateral respectively longitudinal main orientation, a relatively sharp transition zone exists where the main fibre orientation changes between these two extremes.

The centre plot of figure 4.17 shows that no fibres are oriented in through-thickness direction.

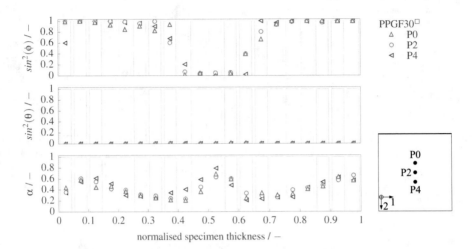

Figure 4.17   Fibre orientation of PPGF30$^\square$ in longitudinal positions.

The lower part of the figure depicts that the total amount of fibres actually oriented in the previously discussed main orientation direction varies over the entire sample thickness. Anisotropy degree α is highest in regions of a distinct longitudinal or lateral fibre orientation, while being slightly less pronounced in transition zones. The most prominent anisotropy degree is found in the core layer. Regarding the three longitudinal evaluation positions, distributions are found to be approximately homogeneous, with slight variations in the transition zone from longitudinal to lateral orientation.

The discussed FOD values support the tensile results from figure 4.16. Stiffness and strength are highest in longitudinal direction since the entirety of boundary layers with fibres aligned in direction of loading is larger than in the core layer with transversely oriented fibres. However, since one third of the fibres is oriented transversely, the contrast of lateral and longitudinal tensile behaviour is not strongly pronounced.

In comparison to the newly proposed visualisation approach applied in figure 4.17, the diagonal elements of the second order orientation tensor in the plate reference system are shown in figure 4.18. As described in section 4.3.3.2, the second order fibre orientation tensor distribution is commonly visualised in current publications by displaying these diagonal elements. The issue concerning the negligence of the remaining tensor components in regions of the part that are not evaluated in the principal axis system has been discussed in section 4.3.3.2. This issue becomes obvious when comparing the graphs of figure 4.17 and 4.18. When lacking the information that the off-diagonal tensor components not given in figure 4.17 may have non-zero values, this representation may lead to the conclusion that the core layer with orthogonal fibres is smaller than it actually is.

Moreover, it may seam that the transition zone between the skin and the core layer is not as distinct as it was presented in 4.17. Next to lacking full information on the orientation tensor components, the graph of figure 4.18 does not allow the differentiation of the principal fibre orientation and the actual degree of anisotropy.

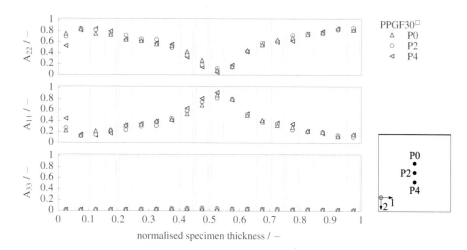

Figure 4.18   Diagonal elements of the second order orientation tensor of PPGF30$^\square$ in longitudinal positions.

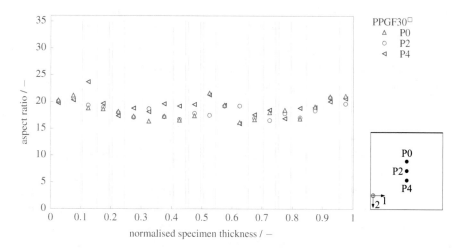

Figure 4.19   Fibre aspect ratio of PPGF30$^\square$ in longitudinal positions.

Fibre aspect ratio was found to be similar for all samples. From figure 4.19 it is evident that the fibre aspect ratio is not constant in through-thickness direction, showing values that deviate approximately ±20 % from average. The variation of the aspect ratio over sample height seems to be typical for injection moulded samples and is thoroughly discussed in MÖNNICH, 2015. Process simulations currently do not take into account the fibre length distribution. Qualitative similarity of the curves in figure 4.19 with the aforementioned shape of the anisotropy degree emphasises that a connection between anisotropy amount and aspect ratio exists. In part, this is a result of the weighted orientation tensor evaluation introduced in equation (2.6).

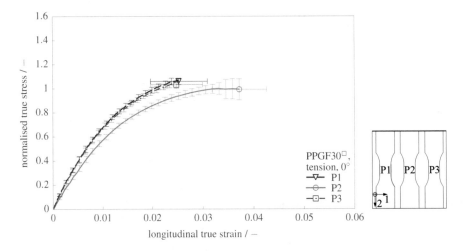

Figure 4.20   Extraction position influence on stress-strain behaviour of PPGF30□.

Considering different lateral extraction positions, the tensile responses of specimens from the PPGF30□ part show noticeable differences, as seen in figure 4.20. Tensile bars from positions P1 and P3 display similar stiffness and strength capabilities, which surpass those found at position P2. Fracture strain at these positions is about 70 % of the maximum strain at position P2.

The mechanical results presented in figure 4.20 correspond to the FOD results for the respective positions given in figure 4.21. It is visible that the principal fibre direction in the core layers at positions P1 and P3 is not completely aligned in lateral direction but shows a diagonal characteristic. This means that a fraction of fibres in the core layers influences load bearing in longitudinal direction. P2 shows a distinct transverse core layer and hence has reduced load bearing capabilities in longitudinal direction. Anisotropy degrees in figure 4.21 show a similar variation at all positions. The reason for the fibre orientation characteristics at positions P1 and P3 in the plate was not evaluated in detail in this work. It is most likely accountable to the form of the melt front during injection moulding.

Besides testing of the 80 mm × 80 mm × 2.5 mm plate (□), additionally the injection moulded flat bar ▯ was evaluated.

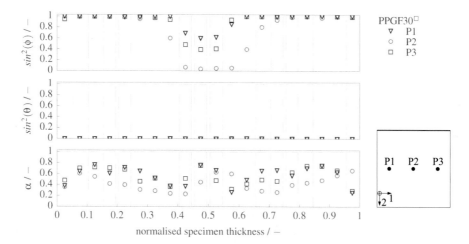

Figure 4.21   Fibre orientation of PPGF30$^\square$ in lateral positions.

As described in section 4.3.1.2 a small tension bar was applied in order to make the evaluation of the transverse properties of the flat bar specimen possible. This small tension bar was also used to derive the tensile properties of the PPGF30$^\square$ part in 0°. Thus, it is ensured that mechanical tests results in lateral and longitudinal orientations can be related to the same relevant material volume with its specific fibre orientation distribution. The performance of the small tension bar is checked against the results of a full scale tension bar in figure 4.22 at a testing speed of 1 mm/min. Stress-strain behaviour is similar in the linear elastic region. With ongoing deformation, the small tension bar shows increasingly larger stresses than the standard size tension bar. The stress maximum of the small specimen is approximately 7 % higher than the results from the standard size tension bar. Overall, this behaviour can be ascribed to a higher local strain rate and a more pronounced fibre orientation in the constrained volume of the small tensile bar in comparison to the larger volume of the standard specimen. The mean failure strain evaluated with the small specimen is approximately 15 % higher than the value evaluated with the standard size tension bar. Nevertheless, it lies in the range of the scatter of the larger specimen.

The results from tensile specimens extracted from the flat bar in lateral and longitudinal orientations are presented next to the previously discussed plate behaviour in figure 4.23.

In order to understand the mechanical response of the flat bar, the corresponding FOD is shown in figure 4.24. Figure 4.25 displays the related aspect ratio distribution. The flat bar section at position P2 of the flat bar source part shows a high fibre orientation in main flow direction over the entire specimen thickness. A distinct transverse core layer is absent in this sample. This leads to a stress-strain curve in longitudinal loading direction that significantly exceeds the values determined from plate parts, as seen in figure 4.23.

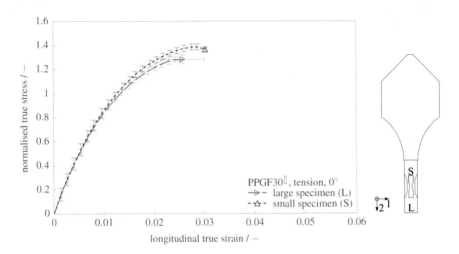

Figure 4.22 Comparison of differently sized tensile samples extracted from PPGF30[l].

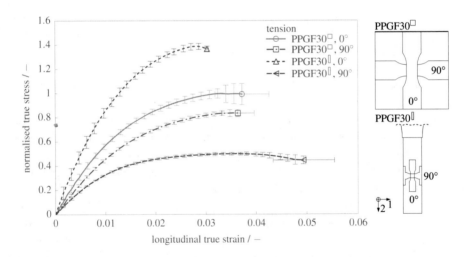

Figure 4.23 Source part influence on stress-strain behaviour of PPGF30.

Lateral tensile specimens from the flat bar display decreased stiffness and strength in comparison to the samples from the plate. Flat bar samples show a large deformation capability in 90° direction and much more brittle behaviour at 0°. Failure strain dependency on loading direction is less distinct in specimens extracted from plate parts.

As shown in figure 4.17 and 4.24, the FOD is of similar form in the plate section of the flat bar part at position P0 in comparison to the distribution found in the standard plate part. As in the plate part, a pronounced transverse core layer can be found at this position of the

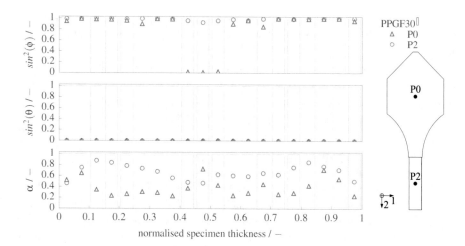

Figure 4.24   Fibre orientation of PPGF30⊔ in flat bar source parts.

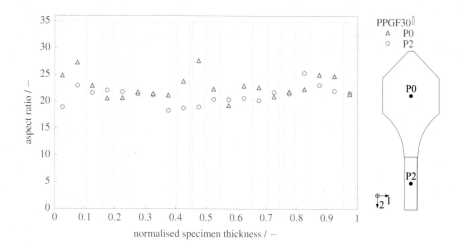

Figure 4.25   Fibre aspect ratio in PPGF30⊔.

flat bar source part. Overall, the anisotropy degree α shows a less pronounced anisotropy in the flat bar at P0 than in the plate source part.

The evaluation of the FAD in figure 4.25 shows that the aspect ratio is slightly reduced in the flat bar area at the end of flow region in comparison to the plate section near to the gate. The most notable difference is that the aspect ratio is larger in the core section of the specimen at position P0.

Results so far emphasise that micro-mechanical conditions have to be considered when predicting material behaviour. A simple assumption of longitudinal samples giving the strongest mechanical response and lateral samples giving the weakest, while values in all other loading directions can be interpolated to lie between those extremes, will lead to an inadequate approximation of material behaviour. Local anisotropy needs to be considered regarding loading direction and loading position. Furthermore, the anisotropy depends on the injection moulded part the sample has been extracted from.

4.4.2 Strain Rate Dependent Behaviour

Figure 4.26 depicts behaviour of PP$^{\square}$ at high deformation rates. The data evaluated on the high speed testing machine shows higher material strength when strain rate is increased. This effect is most significant between quasi-static and low velocity loading and less pronounced between curves at higher strain rates. Next to higher strength values, a considerable loss of ductility is depicted in figure 4.26. Strain rate increase leads to a brittle material response. The effects of strain rate dependent strength increase and fracture strain reduction are typically found for non-reinforced thermoplastic materials, compare e.g. DOMININGHAUS, 2008 and BECKER, 2009. The decrease of fracture strain is often less pronounced for other thermoplastic matrix materials than it was found for the evaluated PP material.

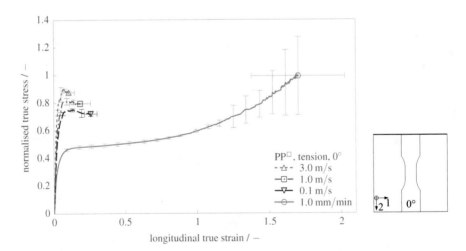

Figure 4.26 Strain rate dependency of PP$^{\square}$ on overall stress-strain behaviour. Data was normalised to the maximum stress value of the quasi-static curve.

The reinforced PPGF30$^{\square}$ was subjected to high speed tests at loading directions of 0°, 90° and 45°. Results of these tests are summarised in figure 4.27, figure 4.28 and figure 4.29.

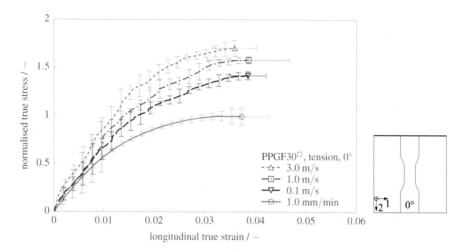

Figure 4.27 Strain rate dependency on stress-strain behaviour of tensile PPGF30$^\square$ samples extracted in 0°.

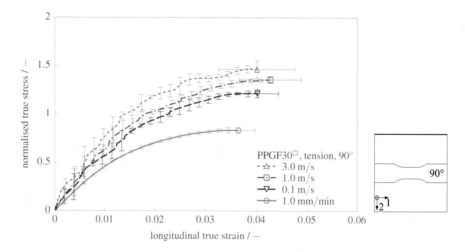

Figure 4.28 Strain rate dependency on stress-strain behaviour of tensile PPGF30$^\square$ samples extracted in 90°.

For all tested loading angles it can be seen that material stresses increase with raised deformation speeds. This essentially corresponds to the behaviour found in the PP$^\square$ samples. Higher strain rates lead to a stiffer elastic behaviour and increased plastic stresses.

It can be perceived in figures 4.27, 4.28 and 4.29 that the scatter of the experiments is significantly higher at higher strain rates than at quasi-static loading. The reason for this behaviour can be found in the temperature dependent characteristics of the material. These

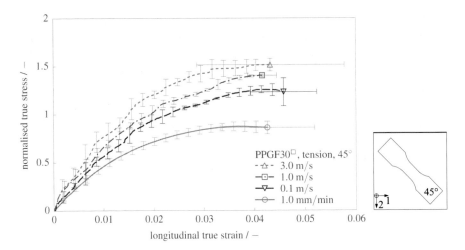

Figure 4.29 Strain rate dependency on stress-strain behaviour of tensile PPGF30$^\square$ samples extracted in 45°.

characteristics have been evaluated by means of dynamic mechanical analysis (DMA) in KEHRER et al., 2015 for the same PPGF30 composite and PP matrix material used in the present work. It was evaluated that the glass transition temperature of PP is approximately at 7 °C. The results found by KEHRER et al., 2015 are presented in section 4.5. The transition range of the PPGF30 composite encompasses the region of room temperature. This means that the influence of a temperature change in this region has a large impact on mechanical material characteristics. It is shown by e.g. RÜHL, 2017 and KUNKEL, 2018 that plastic materials show adiabatic heating at high loading velocities. It can be considered that such an effect also occurs during the presented experiments with PPGF30 at higher strain rates. However, this was not explicitly evaluated experimentally and the measured stress-strain data gives implicit material properties regarding the coupled effect of adiabatic heating influences and strain rate dependency. In summary, inhomogeneous temperature distributions during the tests may lead to the high scatter of mechanical properties due to the temperature range of the glass transition region.

Unlike PP$^\square$, failure strains of PPGF30$^\square$ seem to be depending on neither strain rate nor loading orientation. Regarding the strong dependency found in PP$^\square$ such a behaviour of PPGF30$^\square$ is physically only explicable if the interface between the fibres and the matrix is not ideal. It is therefore assumed that bonding of fibres and the polymeric matrix is insufficient and the interface is prone to failure, hereby influencing overall material reactions. In order to further inspect this effect, fracture surfaces of the tensile bars were inspected by means of SEM.

An SEM micrograph of an exemplary fracture surface is shown in figure 4.30. The qualitative SEM failure surface analysis substantiates the assumption of a weak interface

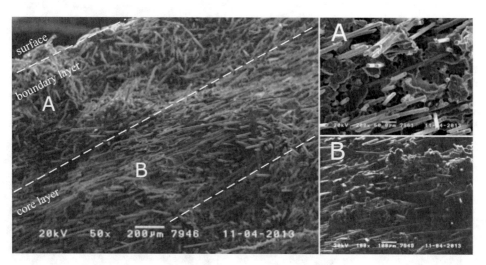

Figure 4.30 SEM analysis of PPGF30$^\square$ tensile bar fracture surfaces.

between the fibres and the matrix. Region A in figure 4.30 represents a skin zone with a mostly longitudinal fibre distribution. Here a high fraction of pulled out fibres and related voids in the matrix material is visible. The pulled out fibres protruding from the fracture surface are clear of matrix residue. This leads to the conclusion that bonding between the fibres and the matrix was weak. In case of a strong interface, rather the matrix would have fractured and corresponding remains would have stuck to the fibre ends. Deformation and failure of the matrix material can be seen to occur in core sample regions with a mostly lateral fibre distribution, as seen in figure 4.30 area B. This becomes clear from the remains of plastically deformed matrix fractions protruding form the failure surface.

Interface phases of the self-compounded material PPGF30 could have been improved by an adjustment of coupling agents. However, an iterative improvement of the compound was not perceived to be expedient in the context of this work.

The large amount of stress-strain data acquired in figures 4.27, 4.28 and 4.29 shows qualitative similarities regarding strain rate dependent behaviour at different loading angles. In order to quantify these similarities, in a first step the first maximum stress values σ_M of all curves in each loading direction were normalised in respect to maximum stress of the related quasi-static curve $\sigma_{M,\dot{\varepsilon}_{QS}}$. The values were then mapped to the corresponding average strain rate during respective experiments. The results are presented in figure 4.31.

Due to normalisation, maximum quasi-static stress values have the value unity on the y-axis. The normalised maximum stress values at higher strain rates all lie in close proximity to one another, seemingly regardless of loading direction. The data points determined from pure PP$^\square$ behaviour deviate from the PPGF30$^\square$ values. Results suggest that PPGF30$^\square$ strain rate dependency can be approximated independently of orientation considerations by

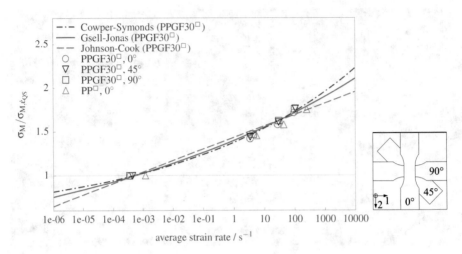

Figure 4.31   Analysis of first maximum tensile stress values σ_M of PPGF30$^{\square}$ at average strain rates in relation to quasi-static response.

a scaling of quasi-static stress-strain curves. The Johnson-Cook, G'Sell-Jonas and Cowper-Symonds scaling approximations given in chapter 4.3.2.6 were fitted to the PPGF30$^{\square}$ points of figure 4.31 averaged over all orientations. The respective parameters are given in table 4.1. The value for the reference strain $\dot{\varepsilon}_0$ of the approximations has been set constant. Nonetheless, the value for the reference stress σ_0 was used as a fitting parameter and not set constant in order to achieve the best possible fit for all strain rates.

Table 4.1   Parameters for the strain rate fitting of average normalised maximum stress values of PPGF30$^{\square}$ and PP$^{\square}$.

Fit	$\dot{\varepsilon}_0$	σ_0	ζ
G'Sell-Jonas	0.00037	0.9837	0.044545
Johnson-Cook	0.00037	0.9827	0.133255
Cowper-Symonds	0.00037	0.9890	13.640977

The curves suggest that a single parameter set derived from results in one loading direction can be used to approximate behaviour of samples in arbitrary orientations. In a next step, this assumption was further investigated by an analysis of the entire plastic deformation range. From each of the stress-strain curves in figures 4.27, 4.28 and 4.29 the plastic stress response at a constant strain rate was extracted.

Behaviour at constant strain rates was determined by use of the method described in chapter 4.3.2.6. At first a 3D surface was fitted to the curves using the G'Sell-Jonas expression in (4.9). The fit was applied with 100 strain increments. The fit parameters from 10 of these points are given in tables 4.2 and 4.3. For the 0° loading direction the resulting 3D fit surface is shown in figure 4.33. Stress-strain curves were extracted from the 3D surface at the constant strain rates 0.0004/s, 3/s, 30/s and 100/s. These values approximately correspond to the average strain rates of the tensile experimental data evaluated at machine velocities of 1 mm/min, 0.1 m/s, 1 m/s and 3 m/s.

Plastic stress curves were derived from the stress-strain response at constant strain rates by the method shown in chapter 4.3.2.4.

The plastic stress values for all constant strain rates where then normalised with respect to the quasi-static stress curve $\sigma_{\dot{\varepsilon}_{QS}}$ for each loading direction. The final results of this procedure are given in figure 4.32. The normalised results of figure 4.32 and the fit parameters given in tables 4.2 and 4.3 show that the strain rate dependency varies with increasing plastic strain. Overall, the strain rate influence decreases at higher plastic strains.

Normalised plastic deformation curves of PPGF30$^{\square}$ at equal strain rates lie approximately in the same range, regardless of loading direction. Essentially, the previous conclusion is confirmed by figure 4.32, that a single parameter set for one loading direction can be used to approximate the dynamic behaviour of samples in arbitrary orientations. This assumption seems to hold for the entire plastic deformation range.

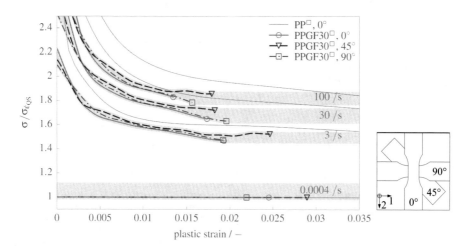

Figure 4.32 Analysis of tensile stress values of PPGF30$^{\square}$ at constant strain rate in relation to quasi-static response.

The values of PP$^{\square}$ differ slightly. This basically means that the fibre volume fraction has an influence on strain rate dependency. Such has to be expected since strain rate dependency must lie between the extremes of pure PP in case of a non-existent fibre reinforcement and pure glass in case of a matrix fraction of zero.

This leads to the result that strain rate dependency can be approximated to be independent of reinforcement fibres, regarding the main orientation as well as the fibre orientation distribution, at least at the fibre volume fractions evaluated in the context of this work. This effect can be attributed to the fact that the SFRTP strain rate dependency is matrix-driven.

SCHÖPFER, 2011 also found that the short glass fibre reinforced polyamide matrix composite considered in his work did not show a pronounced anisotropy with respect to viscoplasticity similar to the present PPGF30 material. Hence, anisotropy regarding strain rate dependency was neglected in the modelling approach made by VOGLER, 2014.

For this reason, a transferability of the results found by the presented characterisation of the PP matrix composite, to short fibre reinforced composites with differing thermoplastic matrices could be expected. However, the range of validity of this assumption was not further inspected in the present work apart from the PPGF30 composite. It will have to be investigated for different matrix materials as well as deviating volume fractions in further research. Some considerations regarding the validity of the assumption based on the micro-mechanical model are given in section 5.3.10.

Table 4.2   Parameters for the 3D surface with a G'Sell-Jonas fit at 10 strain points for 0° loading.

Loading direction 0°		
Strain	**Fit parameters**	
	$\dot{\varepsilon}_0 = 0.00039$	
ε_n	σ_0	ζ^{GJ}
2.5e-05	0.121	0.08971
0.004364	17.971	0.02204
0.008782	32.437	0.02882
0.013130	43.196	0.03348
0.017533	50.265	0.03662
0.021700	55.006	0.03984
0.025838	58.731	0.04054
0.029639	61.316	0.04063
0.033920	61.324	0.04426
0.038724	60.407	0.04601

Table 4.3 Parameters for the 3D surface with a G'Sell-Jonas fit at 10 strain points for 45° and 90° loading.

Loading direction 45°			Loading direction 90°		
Strain	**Fit parameters**		**Strain**	**Fit parameters**	
	$\dot{\varepsilon}_0 = 0.00048$			$\dot{\varepsilon}_0 = 0.00036$	
ε_n	σ_0	ζ^{GJ}	ε_n	σ_0	ζ^{GJ}
2.5e-05	0.050	0.16426	2.5e-05	0.074	0.11521
0.005159	16.340	0.03731	0.004828	15.690	0.03201
0.010385	29.372	0.03886	0.009718	27.904	0.03594
0.015528	39.034	0.04015	0.014529	37.422	0.03834
0.020736	45.975	0.04002	0.019402	43.758	0.03905
0.025666	50.007	0.04140	0.024015	48.095	0.03947
0.030561	52.999	0.04161	0.028595	50.959	0.04111
0.035057	54.548	0.04302	0.032801	53.192	0.04067
0.040122	52.330	0.05047	0.037540	53.273	0.04357
0.045805	53.220	0.04991	0.042857	53.262	0.04383

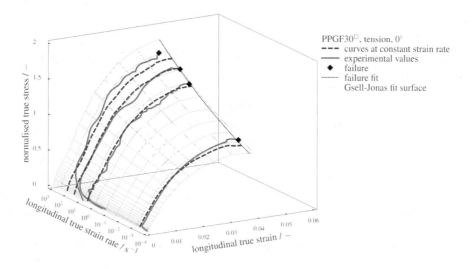

Figure 4.33 3D surface for the extraction of constant strain rate curves of PPGF30□ loaded in 0°.

4.4.3 Damage Behaviour

In order to specify material damage during uniaxial loading, cyclic tensile tests were performed at quasi-static loading rates. The samples were strained until a predefined total deformation was reached. Thereafter, the external load was removed. The samples contracted elastically until a global stress value of zero was attained and mainly plastic strains remained. In each subsequent loading step, deformation was increased in relation to the previous loading cycle. This procedure was repeated until the failure strain was reached.

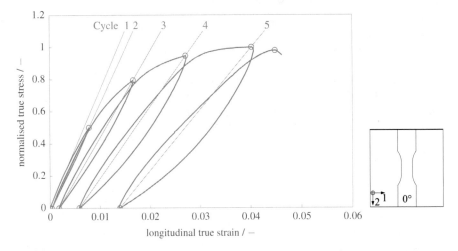

Figure 4.34 Exemplary cyclic loading of PPGF30□ in 0°. The secant tensile modulus for each cycle is presented by a solid line.

An exemplary stress-strain response of a cyclic deformation test is displayed in figure 4.34. It can be seen that the modulus of the sample reduces with each loading step. As described in chapter 4.3.2.5 this could be ascribed to a reduction of the effective cross section due to arising micro-cracks and voids. Actual material modulus remains constant during loading. Because of viscous effects, material response during elastic deformation is not linear. In order to evaluate the decrease of modulus, a linear approximation of elastic material behaviour is presented in figure 4.34. The initial material modulus is evaluated by the method described in section 4.3.2.4. For the modulus evaluation of the preloaded samples, a secant approximation is applied. Viscous effects are neglected. The secant approximation ensures that the experimental stress and strain values at the beginning and ending of each cycle are met.

Cyclic tensile tests were performed on specimens extracted from plate source parts in different orientations. The results of all of these experiments with the PPGF30□ samples are summarised in figure 4.35 where the remaining modulus is plotted against remaining plastic strain after unloading.

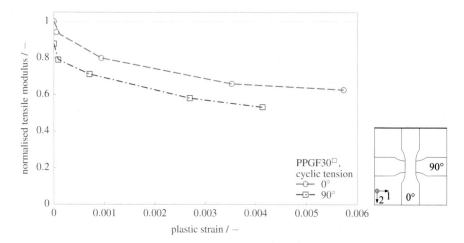

Figure 4.35   Reduction of Young's modulus of PPGF30$^\square$ upon cyclic loading. The data is normalised in respect to the initial Young's modulus of PPGF30$^\square$ loaded in 0°.

The results for different orientations in figure 4.35 show qualitatively similar behaviour regarding the decrease of tensile modulus during cyclic loading. In order to further specify this qualitative similarity, each of the curves is normalised in respect to the material modulus, respectively initial undamaged modulus. The results of this procedure are depicted in figure 4.36.

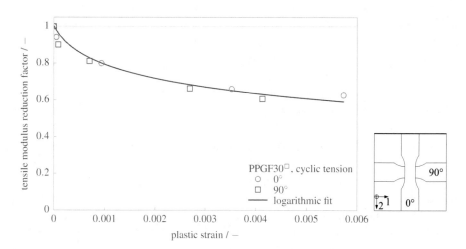

Figure 4.36   Reduction of normalised tensile modulus of PPGF30$^\square$ upon cyclic loading.

The data points from different samples given in figure 4.36 at different deformation steps all lie in a similar range. As a result, it is assumed that the effect of anisotropy on damage behaviour can be neglected.

The damage behaviour of all samples tested can be approximated by a single parameter set. This is shown in figure 4.36. The normalised modulus reduction factors plotted on the y-axis of the figure represent the value $(1 - d)$ given in expression (3.61). The experimentally acquired damage variable d is fitted by the logarithmic expression

$$d(\varepsilon^{\mathrm{pl}}) = \tilde{\zeta}_0 \ln\left(\varepsilon^{\mathrm{pl}}\tilde{\zeta}_1 + 1\right). \tag{4.15}$$

The parameters for the fit shown in figure 4.36 are given in table 4.4.

Table 4.4 Parameters for the fitting of the damage variable.

$\tilde{\zeta}_0$	$\tilde{\zeta}_1$
0.129 89	3874.849

Influence of anisotropy on damage evolution during plastic loading seems to be negligible. At this point it is not clear if this can be purely ascribed to matrix behaviour or if there is a superposition of effects of a matrix and a fibre degradation. For this reason, the behaviour of plastic loading on the fibre distribution has to be evaluated. FODs could not be determined during cyclic loading of the aforementioned samples. This would have made it necessary to perform the tests inside a μCT set-up for in-situ structural analysis, which was not possible during this work.

Alternatively, incrementally loaded samples were tested. Tensile specimens were loaded until a certain deformation state was reached and they were then removed from the testing machine. Exemplary stress-strain results of these tests are shown in figure 4.37. State 0 represents the unloaded sample. At state 1 a plastic strain of approximately 0.0014 remained after unloading. In this case, the remaining strain was measured 10 min after unloading, to account for viscoelastic effects. The last state that was taken into account was the fractured specimen. At this state the μCT analysis was performed in a region next to the fracture surface.

μCT samples were extracted from the pre-strained tensile specimens. Results of the fibre analysis on samples from each deformation state are presented in figure 4.38. Regarding the main planar and spatial fibre orientation as well as the degree of anisotropy α over the entire sample thickness, no distinct differences can be seen between subsequent loading steps. Table 4.5 shows the relative difference of the mean fibre aspect ratio for different incremental loading states in comparison to the unloaded state. Similarly to the FOD, no distinct effect of loading can be seen. The perceivable scatter of the FODs and the deviation of values of the aspect ratios is most probably a result of statistical measurement uncertainty,

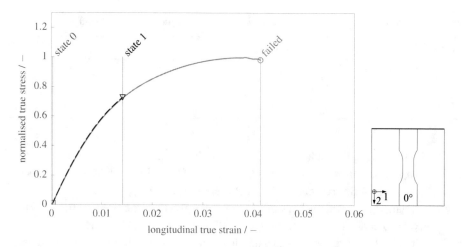

Figure 4.37 Exemplary incremental tensile loading of PPGF30$^\square$ 0° samples for the extraction of µCT specimens.

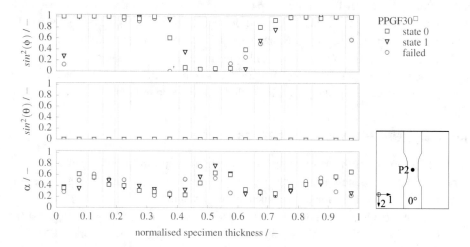

Figure 4.38 Fibre orientation of PPGF30$^\square$ tensile bars at different deformation states.

since a different sample had to be evaluated for each loading step. Furthermore, only a single sample was analysed for each loading step.

In conclusion these results indicate that during deformation no relevant re-orientation of fibres as well as only an irrelevant, if any, breakage of fibres occurs. Nevertheless, this should be evaluated by direct µCT evaluation during loading in future research.

Damage effects are therefore not ascribed to a manipulation of reinforcement fibres. This assumption was checked for different orientation distributions.

Table 4.5 Relative difference of aspect ratio of μCT samples extracted from PPGF30$^\square$.

Sample	Relative difference of aspect ratio
PPGF30$^\square$, P2, state 0	$\frac{\varphi_{\text{state}0} - \varphi_{\text{state}0}}{\varphi_{\text{state}0}} = 0.0$
PPGF30$^\square$, P2, state 1	$\frac{\varphi_{\text{state}1} - \varphi_{\text{state}0}}{\varphi_{\text{state}0}} = -0.012$
PPGF30$^\square$, P2, failed	$\frac{\varphi_{\text{failed}} - \varphi_{\text{state}0}}{\varphi_{\text{state}0}} = 0.036$

Besides incrementally loading tensile samples, shear tests were performed at different deformation states. Here again, no distinct change of FOD or FAD could be determined.

In-situ analysis of SAITO et al., 2017 for PA reinforced with low amounts of fibres have shown that damage and fracture resulted from void growth at the fibre-matrix interface. However, SAITO et al., 2017 perceived a reorientation of fibres during loading. This could be ascribed to the low fibre content and a missing fibre-fibre interaction. Multiple studies, as performed e.g. by FLIEGENER et al., 2017, NCIRI, 2017, SCHAAF, 2015, have shown that in case of composites with short fibre volume fractions comparable to this work, damage and fracture can be ascribed to the matrix and the fibre-matrix interface, whereas fibre fracture will generally not occur. This coincides with the findings of the present work. In summary, the prior assumption holds that damage behaviour is matrix-driven.

Overall, the findings in this chapter only give an overview and and an approximative description of the effect of damage. A more detailed inspection of damage mechanisms based on a larger experimental program should be considered in future research.

4.4.4 Dependency on the Stress State

Figure 4.39 shows a comparison of tensile, compression and shear tests from specimens extracted at 0° from plate source parts. The von Mises stresses at different stress states are plotted as a function of the main component of the strain tensor.

The results are typical for a thermoplastic material, with compressive strength and shear strength being higher than tensile strength at low deformations, cf. for example BECKER, 2009. With increasing deformation the hardened tensile stress-strain curve surpasses those in shear and compression. Shear failure strain is significantly lower than tensile failure strain. During compression, the samples reach highly compressed states and no material fracture was explicitly detected.

Tension, compression and shear tests were also performed with the reinforced PPGF30 material. Results are given in figure 4.40. While the relations between stress values of different stress states may be different in regard to the pure matrix behaviour, it can still be seen that a variance of stress state will lead to a different stress-strain response.

When considering the tested SFRTP, maximum stresses under shear are lower than those under tension, while compression testing leads to increased stress response. In comparison

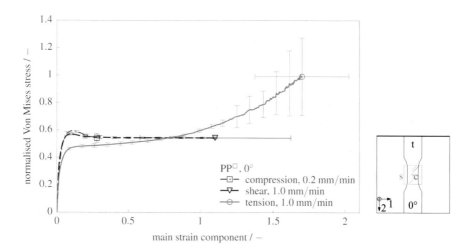

Figure 4.39   Results from tensile, shear and compression test of PP$^\square$ at 0°.

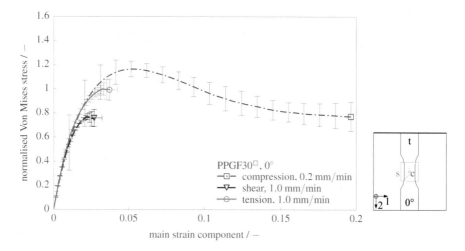

Figure 4.40   Results from tensile, shear and compression test of PPGF30$^\square$ at 0°.

to the pure matrix behaviour, it is visible that the addition of short fibres leads to a significant decrease of shear loading capability in comparison to tensile loading. Still, regarding the maximum stress response, the relation of compressive to tensile behaviour seems not to be effected by the addition of fibres. Overall, the asymmetric behaviour of SFRTP, in respect to the type of loading, must be considered during material modelling.

In literature scaling factors can be found to derive maximum stress in compression and shear from tensile tests. E.g. NUTINI et al., 2017 list a scaling factor of 1.3 for compression and value of 0.8 for shear of a different PPGF30 material. It can be seen in 4.40 that the maximum von Mises stress of the shear test of the present work is related to the maximum tensile stress by the factor 0.78. Thus, it closely matches the value found in NUTINI et al., 2017. However, the maximum stress relation of compression to tension found in the experiments is 1.17. This shows that the relation of these quantities is highly material dependent and should not be generalised.

The fracture strain of PPGF30$^\square$ is lowest in shear loading in comparison to tensile and compressive loading.

Figure 4.40 shows that the stiffness response of the different samples is in a similar range, emphasizing the validity of the comparison. This is due to the fact, that all samples were extracted from the same source part. If the samples had been obtained from different source parts, the effect of alternate fibre orientations would have superimposed the effect of different loading types. The stiffness behaviour under different loading types would not coincide, making it challenging to distinguish the two effects, compare for example KRIVACHY, 2007.

The compression and shear experiments were performed with samples extracted in different orientations.

Results for shear loading are presented in figure 4.41. 0° and 90° give comparable shear values. The 45° samples show an increased stiffness response and a smaller fracture strain.

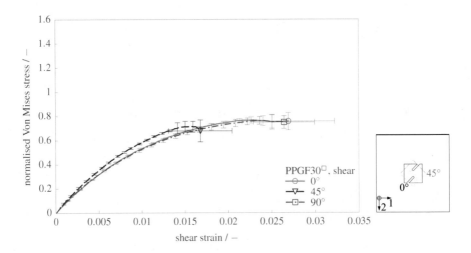

Figure 4.41   Results from shear test of PPGF30$^\square$.

Stress-strain results derived from compression tests are displayed in figure 4.42 for samples from PPGF30$^\square$ and in figure 4.42 for samples from PPGF30^0. As discussed for PP$^\square$, the end of the curves does not represent a point of material fracture. It rather shows the end of the evaluable strain range. After these strain values, the compression of the samples is too high to allow any strain evaluation. Compression experiments show decreased stiffness and strength when loaded transversely to the direction of flow.

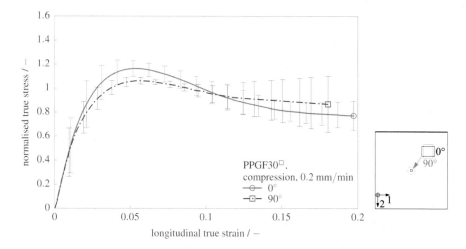

Figure 4.42 Results from compression testing of PPGF30$^\square$.

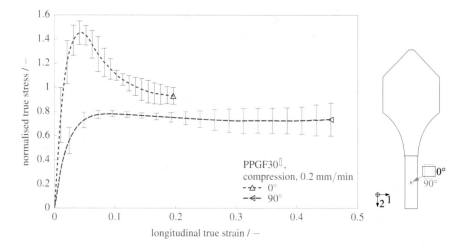

Figure 4.43 Results from compression testing of PPGF30$^\parallel$.

The high scatter of results during compression can be ascribed to effects of friction and non-homogeneous deformation. Apart from PPGF30$^\parallel$ samples loaded in 90°, the stress-strain curves show a pronounced decrease of stress after a stress maximum is reached. Up to the point of maximum stress the deformation of the samples is perceived to be homogeneous. When this point is surpassed the deformation of the samples shows largely non-homogeneous behaviour. In this case, a non-uniaxial stress state has to be considered. Merely the 90° PPGF30□ samples show a mainly homogeneous deformation behaviour throughout the entire deformation process. These effects are discussed in detail in the following.

Compression samples for non-reinforced PP□ give reasonable results over a large deformation range. When friction reduction methods are applied, these samples show an approximately homogeneous straining without bulging, as is seen in figure 4.44 and is discussed in DILLENBERGER, 2011 and DILLENBERGER, 2014. In contrast, the PPGF30□ specimens, from which curves of figure 4.42 were derived, show distinct bulging effects at large deformations. At the beginning of loading, the samples depict homogeneous behaviour. With continued straining, deformation becomes increasingly non-homogeneous and instabilities occur. Figure 4.44 shows a comparison of PP and PPGF30 samples at higher deformation where the non-homogeneous deformation of the PPGF30 sample is clearly visible.

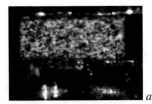 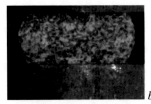

a   *b*

Figure 4.44 Compression samples: *(a)* homogeneous deformation of PP□ and *(b)* non-homogeneous deformation of PPGF30□.

An interesting aspect of compression tests of SFRTP samples is that the individual fibre orientation zones show a different behaviour under compressive loading.

In the following, merely the phenomenological manifestation of non-homogeneous deformation under compressive loading is presented. The underlying mechanisms which cause the non-homogeneous bulging deformation of samples have not been evaluated and are topic of further research. The instabilities could be an effect of buckling of fibres, of multiaxial compression stress states or superimposed effects. In this context, the analysis of samples under loading in an in-situ μCT apparatus would be a suitable measure to clarify the influence of fibre buckling.

The skin-core morphology of fibres shown in figure 2.7 becomes evident when regarding the deformation of compression samples taken from plate source parts. Bulging of samples in regions of fibres oriented in loading direction can be experienced while the lateral fibre orientation zones show an increased deformation.

Figure 4.45 shows the deformation of a PPGF30□ compression sample that was loaded in 0°. In this case, boundary layer fibres are oriented in the load path, while core layer fibres show transverse orientation. For this reason, stiffness in the boundary layers is higher than in the core zone in respect to the loading direction. The boundary zones display less plastic deformation, depicted by less compression after load removal. Moreover, bulging of the outer layers occurs. The centre of the sample shows largest deformations in load direction. The deformation of the sample in width direction is significantly higher than in through-thickness direction. Poisson's ratio is lower in through-thickness direction in comparison to width direction, since strains transverse to fibres are larger than in longitudinal orientation. Since straining longitudinal to fibres is restricted in comparison to transverse straining, an indentation in through-thickness direction is perceivable in the core layer of the specimen.

Similar effects are observable in PPGF30$^\square$ specimens loaded in 90° given in figure 4.46. Compared to 0° specimens, the different zones are switched, since the fibre orientation in regard to loading direction is inverted. In this case, non-homogeneous deformation and bulging is perceivable in the core layer.

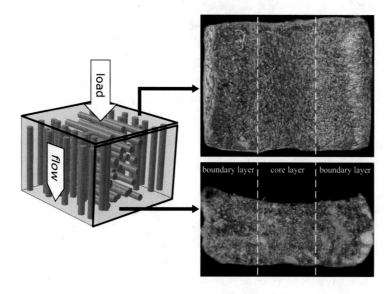

Figure 4.45 Exemplary deformation of a 0° compression sample from PPGF30$^\square$.

Samples extracted from flat bar source parts do not have the distinct core layer with transversely oriented fibres, as was already discussed in chapter 4.4.1. This is why respective compression specimens show deformation patterns that differ from those previously discussed.

Samples loaded in 0° exhibit significant non-homogeneous deformation, as seen in figure 4.47. In this case, all fibres are mostly oriented in the loading path. This leads to high resistance against longitudinal straining. As soon as static friction is surpassed on one surface of the sample, it shows larger transverse deformations in comparison to the opposite loaded surface.

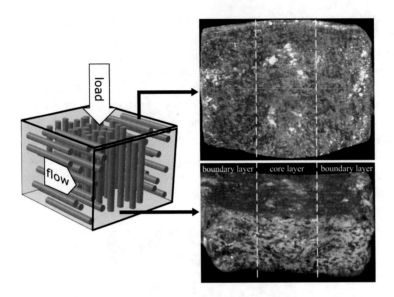

Figure 4.46   Exemplary deformation of a 90° compression sample from PPGF30.

A possible explanation for this behaviour could be that fibres at this side rotate out of the loading path. Following this assumption, stability failure leads to higher straining possibilities on the respective flank of the specimen, since fibres now increasingly orient transversely to the loading path. Hence, deformation localises, resulting in the overall prismatic shape of the tested specimen. In order to validate this assumption, an analysis of samples in an in-situ µCT apparatus would be expedient.

Reduced slenderness of testing samples could possibly avoid this unstable compressive failure. However, a further reduction of the already minimal sample geometries would render experimentation impossible. An alternative for future research could be the extraction of bigger samples with reduced slenderness from approximately uniaxial source parts with higher thickness.

In case of 90° loading, the majority of fibres is oriented transversely. Deformation of the sample can be easily achieved in loading direction. Stability failure does not occur and no buckling loads are applied to the fibres. An overall homogeneous deformation takes place. The originally cubic shape will be deformed to a hexahedron with unequal edge length, as shown in figure 4.48. This can be ascribed to differing Poisson's ratios in width and thickness direction. Through-thickness straining capability is restricted by the fibres oriented in this direction, while no fibres restrict width-related straining.

Regarding the previous observations, it can be stated that during the compression of SFRTPs mostly a non-uniaxial stress state predominates as soon as a certain plastic deformation level is reached. This needs to be considered when applying experimental

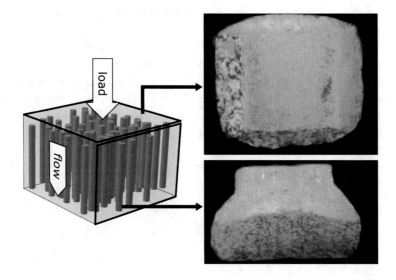

Figure 4.47 Exemplary deformation of a $0°$ compression sample from PPGF30[].

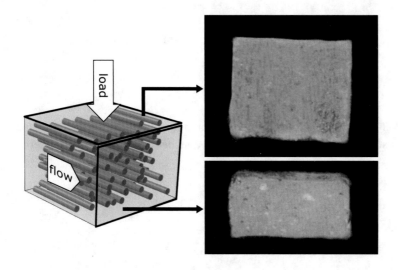

Figure 4.48 Exemplary deformation of a $90°$ compression sample from PPGF30[].

results in material modelling. Merely the transverse test results for approximately uniaxial composites and data from non-reinforced materials can be used without restriction over the entire deformation range.

Next to the tension, compression and shear experiments discussed above, three point bending tests were performed. The results of these tests are presented in figure 4.49. On the one hand, the 0° samples show a distinctly higher load response in comparison to the 90° samples. On the other hand, the samples loaded in 0° show a more brittle behaviour. In this context, the fibre orientation in the skin layers predominates the overall material behaviour. Fibres in the core layer only give a small contribution due to being located around the neutral axis.

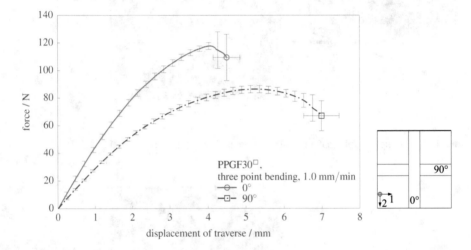

Figure 4.49   Results of three point bending tests with PPGF30□.

The force results from biaxial loading are shown in figure 4.50. The evaluation of the experiments was stopped at the peak force, since it coincided with the point of material fracture.

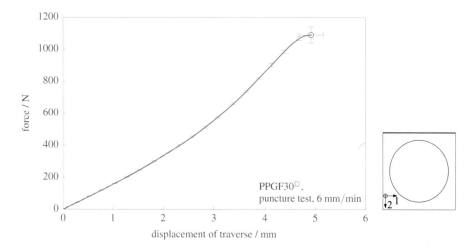

Figure 4.50   Results of biaxial penetration test with PPGF30$^\square$.

4.5 Discussion of Temperature Effects

Herein an outlook is given of further effects that influence the material characteristics. These effects have been neglected and are not considered in the constitutive equations derived in chapter 5, since additional experimental validation would be needed.

Temperature dependent behaviour of the same PPGF30 material that was characterised in the present work has been evaluated by DMA measurements at the Chair for Continuum Mechanics at Karlsruhe Institute of Technology (ITM), cf. KEHRER et al., 2015. The samples consisted of 10 mm × 80 mm × 2.5 mm bars extracted from the PP$^\square$ and the PPGF30$^\square$ parts and were loaded in tensile mode. The DMA was performed with a temperature sweep between −40 °C and 120 °C. Further details on the application and evaluation of DMA tests can be found in e.g. BRYLKA et al., 2018 and RÜHL, 2017.

The results for storage modulus, respectively temperature dependent Young's modulus are given in figure 4.51. The data was normalised to a temperature of 20 °C.

Orientation influence on temperature dependency seems to be negligible for the PPGF30$^\square$ part. The amount of reinforcement fibres has a large influence on overall behaviour as can be seen upon comparison of PPGF30$^\square$ to pure PP$^\square$. In figure 4.51 it is perceivable that influence of the fibre volume fraction on the glass transition region is negligible.

It is shown in figure 4.51 that no evaluation of the PPGF30$^\square$ parts was performed. Furthermore, information on the influences of temperature on the plastic deformation capability and failure of the PPGF30 have not been evaluated. Consequently, the presented data is not applied in the further. It should be considered in future research.

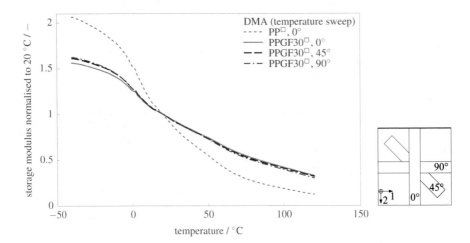

Figure 4.51　Temperature dependency of PPGF30$^{\square}$ on Young's modulus from DMA presented in KEHRER et al., 2015.

5 Material Modelling

5.1 Overview

Based on current approaches of existing material modelling techniques for SFRTPs given in section 2.6, a novel method will be proposed in the present chapter.

Several authors have shown that it is possible to account for the behaviour of SFRTPs by the use of phenomenological models originally developed for metal deep drawing applications. When considering models for sheet metals, the available parameters are used as fitting values in order to achieve an appropriate prediction quality for the polymer materials under consideration. A major drawback of these approaches is the large experimental effort in order to identify the locally varying anisotropic parameters in intricate SFRTP components. The mapping process is elaborate, since these models do not allow a direct input of the fibre orientation. The fibre orientation tensor which is usually determined from an injection moulding simulation has to be adapted to a material orientation for the respective models. Besides depicting main orientation, the second order orientation tensor gives information on the probability of fibre distributions as well. The phenomenological approaches basically only take into account the principal orientation. For this reason, several authors introduce discrete material classes to approximately describe different probability distributions, see e.g. GRUBER et al., 2012, VOGLER, 2014, SCHÖPFER, 2011 and NUTINI, 2010.

As an Alternative to phenomenological methods, micro-mechanical methods can lead to an appropriate prediction of the SFRTP material response. Since micro-mechanical methods essentially take the matrix and the fibre behaviour as input parameters, the amount of experiments needed in order to predict the effective behaviour is reduced. Furthermore, they are capable of handling all aspects of the fibre orientation tensor information directly. Therefore, in contrast to the phenomenological approaches, they can account for behaviour in every orientation configuration and the aforementioned mapping of material classes to discrete orientation distributions is not necessary. A drawback of micro-mechanical models are high computational times for plasticity modelling. Micro-mechanics based plasticity models often lack complex yield surfaces due to complicated numerical derivations of these surfaces, cf. KAISER, 2013. Moreover, it has to be noted that even though the micro-mechanical models claim to be able to derive the overall composite behaviour from the matrix and the fibre information they often lack this capability regarding plasticity prediction. This is due to the fact that, on the one hand, the non-homogeneous stress states cannot be appropriately accounted for by the underlying averaging assumptions and, on the

© Springer Fachmedien Wiesbaden GmbH, part of Springer Nature 2020
F. Dillenberger, *On the anisotropic plastic behaviour of short fibre reinforced thermoplastics and its description by phenomenological material modelling*, Mechanik, Werkstoffe und Konstruktion im Bauwesen 53, https://doi.org/10.1007/978-3-658-28199-1_5

other hand, fibre-matrix interface behaviour is mostly neglected. As a consequence, the matrix material values utilised in these models represent a best-fit to achieve an adequate composite prediction and do not represent actual physical behaviour. Such data is often referred to as virtual matrix information, as described in e.g. PIERARD, 2006, DOGHRI et al., 2006 and KAISER, 2013. For example, KAISER, 2013 has presented a method to overcome the need of virtual matrix data. However, currently only von Mises yield surface definitions can be used for the matrix material in this approach.

A novel material model is derived in the present work, following the previous perceptions on existing modelling approaches and the experimental findings depicted in chapter 4. The basic characteristics of the model are briefly summarised in the following statements.

From a modelling point of view, the new material model provides the subsequent features:

- Characteristics of the phenomenological as well as the micro-mechanical approach are combined.

- The stiffness calculations for linear elastic modelling are performed on the microscale using established micro-mechanical models. Hence, the advantages of micromechanical techniques are maintained on the elastic level.

- The determination of the initial yield surface is derived from a micro-mechanical perspective. The yield surface is then transferred into the macro-scale by applying a phenomenological formulation.

- The plastic deformation behaviour, such as hardening, strain rate dependency, damage and fracture is predicted by a phenomenological approach that is coupled with the aforementioned micro-mechanically determined results by scaling factors. These scaling factors are acquired by fitting the model to experimental data. Thereby, the computational issues of micro-mechanical models regarding plasticity are overcome, while achieving adequate predictions as known from state of the art phenomenological schemes. Overall, the coupling with the micro-mechanical module leads to a simple parametrisation of the plasticity model.

From the perspective of the material characteristics depicted by the experiments in section 4.4, the material model has the following properties:

- Anisotropic elastic behaviour is assumed to be linear and time-independent, i.e. viscoelastic effects are neglected.

- Yield surface and failure are considered to be orthotropic and pressure dependent.

- Strain rate influence is taken into account isotropically. Normalised viscoplastic behaviour is assumed to be independent of loading direction.

- Scalar isotropic damage influence is taken into account. The normalised reduction of stiffness in respect to plastic deformation is considered to be independent of loading direction.

A schematic representation of the previously stated constitutive model aspects is given in figure 5.1 for an exemplary orientation and stress state.

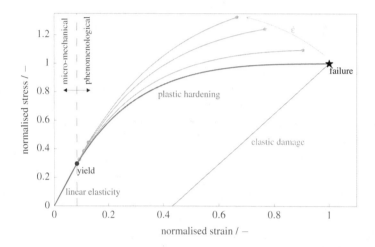

Figure 5.1   Exemplary schematic representation of the aspects of the material modelling approach.

Details on the implemented material model are presented in the next sections.

5.2 Modelling of Linear Elasticity

The following sections briefly highlight the derivation of the micro-mechanical model for the prediction of linear elastic material behaviour. More detailed information on different micro-mechanical approaches can be found in the references presented in section 2.6.

External loading of a homogeneous material results in a load-dependent, homogeneous stress and strain distribution throughout the material volume V. The inclusion of inhomogeneities, such as reinforcement fibres, in such a homogeneous continuum leads to an inhomogeneous stress and strain distribution.

In the following, it is assumed that the total material volume consists of a fibre phase f and a matrix phase m. Furthermore, it is assumed that the volumetric average of the inhomogeneous stresses and strains in the material volume can be described by characteristic volume fractions ϑ of the individual phases and by volumetric stress and strain averages $\langle \boldsymbol{\sigma} \rangle$ and $\langle \boldsymbol{\varepsilon} \rangle$ present in each phase, see e.g. HILL, 1963, BENVENISTE, 1987 and TUCKER et al., 1999. The average stress theorem was postulated by HILL, 1963. Assuming that the volumetric average of the internal stress is equal to the external stresses, it is stated that

$$
\begin{aligned}
\langle\boldsymbol{\sigma}\rangle &= \vartheta^{\mathrm{f}}\langle\boldsymbol{\sigma}^{\mathrm{f}}\rangle + \vartheta^{\mathrm{m}}\langle\boldsymbol{\sigma}^{\mathrm{m}}\rangle \\
&= \vartheta^{\mathrm{f}}\frac{1}{V^{\mathrm{f}}}\int_{V^{\mathrm{f}}}\boldsymbol{\sigma}\,(x)\,dV + \vartheta^{\mathrm{m}}\frac{1}{V^{\mathrm{m}}}\int_{V^{\mathrm{m}}}\boldsymbol{\sigma}\,(x)\,dV, \quad \text{with } \vartheta^{\mathrm{f}} + \vartheta^{\mathrm{m}} = 1.
\end{aligned} \tag{5.1}
$$

This can be equally expressed for average strains as

$$
\begin{aligned}
\langle\boldsymbol{\varepsilon}\rangle &= \vartheta^{\mathrm{f}}\langle\boldsymbol{\varepsilon}^{\mathrm{f}}\rangle + \vartheta^{\mathrm{m}}\langle\boldsymbol{\varepsilon}^{\mathrm{m}}\rangle \\
&= \vartheta^{\mathrm{f}}\frac{1}{V^{\mathrm{f}}}\int_{V^{\mathrm{f}}}\boldsymbol{\varepsilon}\,(x)\,dV + \vartheta^{\mathrm{m}}\frac{1}{V^{\mathrm{m}}}\int_{V^{\mathrm{m}}}\boldsymbol{\varepsilon}\,(x)\,dV.
\end{aligned} \tag{5.2}
$$

In HILL, 1963 a stress concentration tensor $\mathbb{P}$, respectively strain concentration tensor $\mathbb{Q}$ is introduced. These enable the mapping of the volumetric averages of the fibre phase to the volumetric average of the total volume by

$$
\langle\boldsymbol{\sigma}^{\mathrm{f}}\rangle = \mathbb{P} : \langle\boldsymbol{\sigma}\rangle \tag{5.3}
$$

and

$$
\langle\boldsymbol{\varepsilon}^{\mathrm{f}}\rangle = \mathbb{Q} : \langle\boldsymbol{\varepsilon}\rangle. \tag{5.4}
$$

The concept of concentration tensors can be further extended to the mapping of stress, respectively strain response from one material phase to another. In the context of this work, a special stress concentration tensor $\mathbb{T}$ is derived from expressions (5.3) and (5.1) as

$$
\langle\boldsymbol{\sigma}^{\mathrm{m}}\rangle = \mathbb{T} : \langle\boldsymbol{\sigma}\rangle, \quad \text{with } \mathbb{T} = \frac{\boldsymbol{\delta} - \vartheta^{\mathrm{f}}\mathbb{P}}{1 - \vartheta^{\mathrm{f}}}. \tag{5.5}
$$

It allows the mapping of the average total stress to the average matrix stress. In this context, $\boldsymbol{\delta}$ represents the symmetrical unit tensor of fourth order.

By use of certain assumptions, micro-mechanical homogenisation schemes are capable of evaluating average composite properties. These assumptions comprise of an ideal adhesion between the matrix material and the embedded fibres, as well as the knowledge of individual characteristics of the constituents such as stiffness properties, volume fractions and fibre geometries. Using Hooke's law as given in equation (3.35) and combining it with the average stress-strain-postulates from equation (5.1) and (5.2), as well as the respective concentration factors, TUCKER et al., 1999 lists the relationships for the effective stiffness $\mathbb{C}^{\mathrm{UD}}$ respectively compliance $\mathbb{S}^{\mathrm{UD}}$ of a unidirectional composite (UD) as

$$
\mathbb{C}^{\mathrm{UD}} = \mathbb{C}^{\mathrm{m}} + \vartheta^{\mathrm{f}}\left(\mathbb{C}^{\mathrm{f}} - \mathbb{C}^{\mathrm{m}}\right) : \mathbb{Q} \tag{5.6}
$$

and

$$\mathbf{S}^{UD} = \mathbf{S}^m + \vartheta^f \left(\mathbf{S}^f - \mathbf{S}^m \right) : \mathbb{P}. \tag{5.7}$$

Considering Hooke's law as well as equation (5.4) and (5.3), the concentration tensors are related by

$$\mathbb{P} = \left(\mathbf{C}^f : \mathbb{Q} \right) : (\mathbf{C}^{UD})^{-1}. \tag{5.8}$$

Simple bound estimates for the effective composite properties can be derived directly from (5.6) and (5.7), as shown e.g. in TUCKER et al., 1999.

Assuming that the stress concentration tensor equals the symmetrical unit tensor $\boldsymbol{\delta}$, the compliance tensor $\mathbf{S}^{RB}$ for the lower Reuss bound (RB) is established by

$$\mathbf{S}^{RB} = \vartheta^f \mathbf{S}^f - \vartheta^m \mathbf{S}^m \quad (\text{assuming } \mathbb{P} = \boldsymbol{\delta}). \tag{5.9}$$

It can be interpreted as a mechanical series connection of the fibre and matrix phases.

Alternatively the stiffness tensor $\mathbf{C}^{VB}$ of the upper Voigt bound (VB) is recovered by setting the strain concentration tensor equal to the symmetrical unit tensor, resulting in

$$\mathbf{C}^{VB} = \vartheta^f \mathbf{C}^f - \vartheta^m \mathbf{C}^m \quad (\text{assuming } \mathbb{Q} = \boldsymbol{\delta}). \tag{5.10}$$

The Voigt bound considers the composite to be represented by a parallel connection of the fibre and matrix phases.

However, due to describing only special composite configurations, these boundary approaches merely roughly determine the overall composite behaviour. More precise micro-mechanical models make use of the Eshelby solution. ESHELBY, 1957 presented an analytical solution for the strain field of an ellipsoidal inclusion in an infinite matrix. The analysis is performed by virtually subjecting an infinite matrix to a far field strain. After initially considering a homogeneous inclusion to be elastically embedded in the matrix, the strain and stress perturbations resulting from an equivalent non-homogeneous inclusion are evaluated.

The hereby derived Eshelby's tensor $\mathbb{L}$ was later adapted to micro-mechanical problems concerning fibres embedded in a bulk material, by approximating their cylindrical geometry to be ellipsoidal, compare e.g. TUCKER et al., 1999 and GROSS et al., 2011. The ellipsoidal approximation was shown to be adequate for fibres and has led to a multitude of micro-mechanical modelling approaches, capable of improving the aforementioned boundary considerations. As shown in RUSSEL, 1973, results from the Eshelby solution can be directly used to find the strain concentration $\mathbb{Q}^{EY}$ for dilute fibre distributions without fibre interaction as

$$\mathbb{Q}^{EY} = \left[\boldsymbol{\delta} + \mathbf{L} : \mathbb{C}^m : \left(\mathbf{S}^f - \mathbf{S}^m \right) \right]^{-1}. \tag{5.11}$$

Herein, it is assumed that a single fibre is embedded in a bulk material that has the average properties of the matrix.

Eshelby's tensor $\mathbf{L}$ depends solely on stiffness properties of the matrix and the geometrical features of the fibre. The fibre geometry is specified by the aspect ratio φ as introduced in equation (2.8).

Analytical solutions of Eshelby's tensor can be found for isotropic and transversely isotropic matrix materials. These are presented in appendix B.1. Nevertheless, solutions have to be derived numerically for more complex matrix configurations, as shown by e.g. MURA, 1982, WITHERS, 1989 and GAVAZZI et al., 1990. The mechanical properties of the embedded fibres do not influence the Eshelby's tensor. This allows its use for the micro-mechanical modelling of composites reinforced with isotropic glass fibres and transversely isotropic carbon fibres alike. Due to tensor symmetries, the fourth order Eshelby's tensor can be reduced to a matrix system using Voigt's notation given in section 3.3, see e.g. MURA, 1982, TANDON et al., 1984 and KUYKENDALL et al., 2012. The corresponding expressions can be found in chapter B.1. Voigt's notation significantly simplifies the calculations of tensor products and inversions needed for the solution of micro-mechanical models.

Thermoplastic composites in engineering applications usually have fibre mass fractions $>10\%$. The Mori-Tanaka homogenisation model, originally proposed for steel by MORI et al., 1973, has proven to perform well in these applications, cf. section 2.6. Following the interpretation of BENVENISTE, 1987 fibre interactions are indirectly taken into account in the Mori-Tanaka model by imposing that a single fibre in the composite experiences the average strain of the matrix instead of the average deformation of the total composite volume. The resulting strain concentration factor $\mathbb{Q}^{MT}$ for the Mori-Tanaka model is given by

$$\mathbb{Q}^{MT} = \mathbb{Q}^{EY} : \left[\left(1 - \vartheta^f \right) \boldsymbol{\delta} + \vartheta^f \mathbb{Q}^{EY} \right]^{-1}. \tag{5.12}$$

For glass fibre reinforced composites, the Mori-Tanaka model is well established since several authors have shown that it offers accurate predictions for the effective composite properties, see e.g. TUCKER et al., 1999, PIERARD, 2006, KAISER, 2013 and MÜLLER et al., 2015. It can be shown that the Mori-Tanaka model represents a lower bound estimate for the composite stiffness, see e.g. TUCKER et al., 1999. In AMBERG et al., 2013, one of the few experimental validations of micro-mechanical modelling approaches suitable for the prediction of elastic properties of composites with transversely isotropic carbon fibre inclusions can be found. It is shown that carbon fibre reinforced composites can be modelled by the Mori-Tanaka model with reasonable accuracy as well.

The Mori-Tanaka model given by equation (5.12) offers the advantage of being an explicit analytical expression for the calculation of the composite properties. Therefore, it is an attractive tool to numerical applications in engineering where computational times are crucial. An alternative micro-mechanical technique to the Mori-Tanaka model can be found in the self-consistent approach. Using the self-consistent model, the stress concentration tensor has to be calculated implicitly since, in an effort to include fibre interactions, it is assumed that fibres are embedded in a matrix material that has the average properties of the composite and the original stress concentration factor is adjusted accordingly, HILL, 1965a. This leads to an iterative solving scheme for the self-consistent approach which is computationally more costly than the aforementioned explicit Mori-Tanaka model.

While many authors such as e.g. DOGHRI et al., 2006 claim the results of the self-consistent model to be too stiff in comparison to actual experimental data, it has recently been proven that adequate stiffness results are possible by taking into account the actual fibre information gained from composite samples analysed by µCT methods in MÜLLER et al., 2015. However, the results of both the Mori-Tanaka and the self-consistent models are in a similar range as seen in MÜLLER et al., 2015. For this reason, and due to the fact of being computationally more efficient, the Mori-Tanaka approach was applied in this work.

In a computational context, the matrix inversion calculations in (5.12) are numerically demanding. This is why the explicit derivations of the Mori-Tanaka equations given by TANDON et al., 1984 and referred to as Tandon-Weng model are often implemented, see e.g. KAISER, 2013. In order to obtain a model capable of predicting the properties of thermoplastic matrices with transversely isotropic reinforcements such as carbon fibres, the more sophisticated derivation by QUI et al., 1990 can be applied. It originally gives explicit expressions for the properties of composites with multiple transversely isotropic inclusions derived by the Mori-Tanaka approach. In the context of the present work, the Qui-Weng expressions have been adjusted and simplified for the consideration of a single transversely isotropic fibre type. They have been implemented into the micro-mechanical model, see appendix B.2.

Actual injection moulded parts exhibit FLDs and FODs which differ from the unidirectional case. Thus, these distributions must be taken into account in the homogenisation scheme. Mathematically it is possible to incorporate the orientation distribution function directly into expression (5.12). This would make it possible to calculate effective properties of a composite including an FOD directly in one step. Nonetheless, considering pronounced orientation distributions, it has been shown that the results may lack a physical basis, as the direct Mori-Tanaka approach can lead to asymmetric effective stiffness tensors, compare e.g. BENVENISTE, 1987 and QUI et al., 1990.

Accordingly, the alternative two-step orientation averaging scheme has been applied to the calculation of oriented micro-structures in this work.

In the first step, the material volume is divided into volume fractions of equal fibre orientation, the so-called pseudo grains. Each of the pseudo grains is assumed to be of uniaxial nature and the effective properties for a UD material are calculated by the Mori-Tanaka model given in (5.12).

It is possible to consider an FLD in this step of the calculation by regarding each fibre group of equal aspect ratio as a separate inclusion and extend expression (5.12) to multiple inclusions, see e.g. CAMACHO et al., 1990 and KAISER, 2013. Nevertheless, this feature, while having been included in the micro-mechanical model in this work, is omitted in an integrative simulation context, since to date injection moulding simulation tools only offer the calculation of an FOD alongside an average aspect ratio value.

In the second homogenisation step, the evaluated properties of the virtual UD pseudo grains are averaged by a second micro-mechanical model. Such a scheme is described in detail and validated in CAMACHO et al., 1990, DOGHRI et al., 2005, PIERARD, 2006, DRAY et al., 2007, REGNIER et al., 2008 and HESSMAN et al., 2018 amongst others. These authors have shown that the use of Voigt's approximation in the second step, and thus the consideration all pseudo grains being in a mechanically parallel series and experiencing the same strain, gives the best overall prediction for elastic properties. It gives more adequate results than those achieved by the use of Reuss's approximation as well as other micro-mechanical methods. Therefore, in the present work Voigt's bound is applied in the second homogenisation step by

$$\mathbb{C} = \oint \mathbb{C}^{\text{UD}}(\vec{p})\Psi(\vec{p})\,\mathrm{d}\vec{p}. \tag{5.13}$$

This scheme is generally referred to as orientation averaging.

For the case of linear elastic and transversely isotropic pseudo grains, ADVANI et al., 1985 derived an analytical expression for (5.13) using Einstein's summation convention as

$$
\begin{aligned}
C_{ijkl} = {} & A_{ijkl}\left(C^{\text{UD}}_{1111} + C^{\text{UD}}_{2222} - 2C^{\text{UD}}_{1122} - 4C^{\text{UD}}_{1212}\right) \\
& + \left(A_{ij}\delta_{kl} + A_{kl}\delta_{ij}\right)\left(C^{\text{UD}}_{1122} - C^{\text{UD}}_{2233}\right) \\
& + \left(A_{ik}\delta_{jl} + A_{il}\delta_{jk} + A_{jl}\delta_{ik} + A_{jk}\delta_{il}\right)\left(C^{\text{UD}}_{1212} + \frac{1}{2}\left(C^{\text{UD}}_{2233} - C^{\text{UD}}_{2222}\right)\right) \\
& + \left(\delta_{ij}\delta_{kl}\right)C^{\text{UD}}_{2233} \\
& + \left(\delta_{ik}\delta_{jl} + \delta_{il}\delta_{jk}\right)\left(\frac{1}{2}\left(C^{\text{UD}}_{2222} - C^{\text{UD}}_{2233}\right)\right), \quad \text{with } i,j,k,l = 1,2,3.
\end{aligned}
\tag{5.14}
$$

In this context, Einstein's summation convention depicts the summation over repeated indices as defined e.g. in BRONSHTEIN et al., 2007.

This equation makes use of the second order orientation tensor **A** and fourth order orientation tensor **Ⱥ**, which can be computationally handled more effectively than an orientation distribution function, see section 4.3.3.2.

As discussed in section 2.5, injection moulding simulations only offer the output of second order orientation tensors. The fourth order tensor needed for micro-mechanical calculations has to be approximated by closure functions. The adjusted orthotropic fitted closure presented by CHUNG et al., 2001 has been implemented in this work to derive fourth order orientation tensors in a numerical context for the micro-mechanical model. Details on the formulation are given in CHUNG et al., 2001, based on the work of CINTRA et al., 1995.

When using experimental data derived by the μCT fibre image evaluation method presented in GLÖCKNER et al., 2016, closure functions are not necessary since the fourth order orientation tensor can be calculated directly from the spatial distribution of fibres, cf. MÜLLER et al., 2015.

The overall two-step homogenisation scheme, as implemented in this work, is presented in figure 5.2.

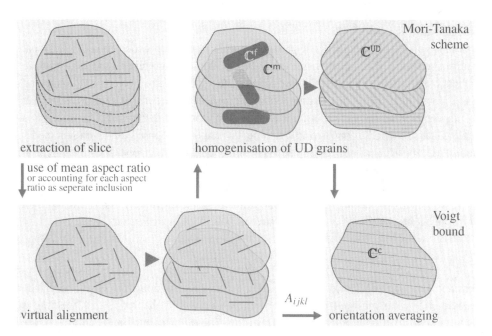

Figure 5.2 Overall homogenisation scheme.

5.3 Modelling of Plastic Behaviour

Micro-mechanical models such as those presented in the previous chapter can also be extended to plastic behaviour. The general implementation of plasticity found in literature follows basically the same procedure as applied to linear elasticity, but it involves several complex extensions due to numerical constraints, as described in e.g. PIERARD, 2006, PIERARD et al., 2007 and KAISER, 2013. Herein, the fibre is approximated to be linearly elastic while only the matrix phase is assumed to behave plastically. Thus, while linear elastic behaviour of the composite is a result of both matrix and fibre properties, the plastic response of the composite can be accounted solely to the matrix.

In case of plastic matrix behaviour, the analytical expressions given for Eshelby's tensor and orientation averaging do not hold. The respective values have to be calculated numerically resulting in a complex computational effort, cf. for example DOGHRI et al., 2006 and KAISER, 2013. The experiments in chapter 4 show that the interface between fibres and matrix is not ideal. This especially influences material failure. Micro-mechanical models incorporating interface failure have been proposed, see for example SCHÖNEICH, 2016. However, they involve even more numerical complexity and have to be extensively calibrated to meet experimental results. This is due to the fact that an experimental derivation of interface properties is not readily performable. Often this issue is overcome by introducing multilevel failure criteria which are calibrated to meet rapture experiments, as done e.g. in KAISER, 2013.

The possibility of calculating the plastic composite behaviour without the computational requirements of the full micro-mechanical model would be preferable in time-sensitive applications. As shown in the following chapters, this issue can be overcome by switching from a micro-mechanical formulation to a phenomenological model in case of plasticity.

Since a detailed inspection of failure was not the topic of the present work, merely a single global failure criterion fitted to experimental results is proposed for the present model. Since plasticity calculation is already done phenomenologically, a global phenomenological fracture criterion can be readily implemented. This follows the experiences from other authors, such as for example SCHÖPFER, 2011, GRUBER et al., 2012 and NUTINI et al., 2017, who have shown that the composite failure can be adequately approximated by a global, phenomenological failure model such as a the Tsai-Wu criterion.

5.3.1 Yield Surface

Thermoplastic materials mostly do not show a distinct transition from elastic to plastic behaviour. For this reason, following the definition given in chapter 4.3.2.4, yielding is defined as the switching point from linear to non-linear behaviour. In the context of the performed experiments, this point was defined by the introduction of an offset yield strength. In the material model the boundary between linear elastic and plastic behaviour is defined by a yield surface. Since the behaviour of the reinforcement fibres is considered to be linear elastic over the entire deformation range, the initiation of non-linear plastic behaviour will be accounted only to the matrix. The yield surface is defined on the matrix level and then transferred to the overall composite.

The definition of an adequate yield surface for the thermoplastic matrix material m is crucial for the subsequent modelling steps. Due to its simple mathematical formulation and the capability of representing relevant differences of tensile, compressive and shear yielding of isotropic polymers, the SAMP model presented in KOLLING et al., 2005 was chosen for this work. The SAMP yield surface formulation given in equation (3.56) is applied to the matrix level of the composite as

$$\left(\sigma^{m,vM}\right)^2 - \xi_0^{SAMP} - \xi_1^{SAMP} p^m - \xi_2^{SAMP} (p^m)^2 \leq 0. \tag{5.15}$$

The parameters ξ_i^{SAMP} (with $i = 1, 2, 3$) of the yield surface have to be derived from the matrix yield values for different stress states. Pressure in the matrix phase p^m is defined in terms of the negative trace of the stress tensor given in equation (3.31). The von Mises stress of the matrix $\sigma^{m,vM}$ is calculated by adopting equation (3.54).

Now it has to be considered how to relate the yield surface on the matrix level to the yield stress of the total composite. At first a transversely isotropic pseudo grain with UD fibres is examined. The designation pseudo grain follows the definition given in section 5.2 for the virtually aligned composite phases defined in the first step of the two-level homogenisation process. Quantities related to a pseudo grain are labelled by the superscript UD in this work. The pseudo grain is subjected to an average stress state $\langle \vec{\sigma}^{UD} \rangle$. Such a stress state is defined by a load stress $\bar{\sigma}^{UD}$ that is distributed onto different stress fractions of the pseudo grain composite by the relation factors $\beta_1, \beta_2, \beta_3, \beta_4, \beta_5$ and β_6, which is expressed as

$$\langle \vec{\sigma}^{UD} \rangle = \bar{\sigma}^{UD} \left[\beta_1, \beta_2, \beta_3, \beta_4, \beta_5, \beta_6\right]^{\mathrm{T}}. \tag{5.16}$$

The SAMP equation can be evaluated on the matrix level for the special load cases of the pseudo grain composite given in expression (5.16) by applying the stress concentration factor $\mathbf{T}$ derived in expression (5.5). For the case of transverse isotropy, $\mathbf{T}$ shows symmetries similar to those of the stiffness tensor discussed in section 3.3. Combining (5.16) and (5.5) and applying these symmetries, the average matrix stress can be derived by

$$\langle \vec{\sigma}^m \rangle = \bar{\sigma}^{UD} \begin{bmatrix} \beta_1 T_{11} + \beta_2 T_{12} + \beta_3 T_{13} \\ \beta_1 T_{21} + \beta_2 T_{22} + \beta_3 T_{23} \\ \beta_1 T_{31} + \beta_2 T_{32} + \beta_3 T_{32} \\ \beta_4 T_{44} \\ \beta_5 T_{55} \\ \beta_6 T_{66} \end{bmatrix} = \bar{\sigma}^{UD} \vec{\sigma}^m. \tag{5.17}$$

In this context, $\vec{\bar{\sigma}}^m$ represents the average matrix stress normalised in respect to $\bar{\sigma}^{UD}$. The normalised matrix stress $\vec{\bar{\sigma}}^m$ can be applied to compute a normalised von Mises stress $\bar{\sigma}^{m,vM}$ and a normalised pressure $\bar{p}^m$ for the matrix by use of equations (3.31) and (3.33). These can be used to express the von Mises stress $\sigma^{m,vM}$ and the pressure p^m of the polymeric matrix material as

$$\sigma^{m,vM}\left(\langle\vec{\bar{\sigma}}^m\rangle\right) = \bar{\sigma}^{UD}\,\bar{\sigma}^{m,vM}$$
$$p^m\left(\langle\vec{\bar{\sigma}}^m\rangle\right) = \bar{\sigma}^{UD}\,\bar{p}^m \tag{5.18}$$

The value $\bar{\sigma}^{UD}$ that occurs when the matrix and thus the composite yields, is designated as σ_Y^{UD}. Inserting the results of equation (5.18) into expression (5.15) and solving it for matrix yield, the composite yield values can be calculated for specific load relation factors by

$$\sigma_Y^{UD} = \frac{-\xi_1^{SAMP}\bar{p}^m \pm \sqrt{-4\xi_0^{SAMP}\xi_2^{SAMP}(\bar{p}^m)^2 + 4\xi_0^{SAMP}\bar{\sigma}^{m,vM} + (\xi_1^{SAMP})^2(\bar{p}^m)^2}}{2\xi_2^{SAMP}(\bar{p}^m)^2 - 2(\bar{\sigma}^{m,vM})^2}. \tag{5.19}$$

The derivation (5.19) applies to a SAMP yield surface formulation for the matrix material. However, the overall procedure presented could be used for arbitrary yield surface expressions.

For metal-matrix composites, the concept of applying stress concentration tensors for the derivation of the composite yield values from a matrix yield surface is applied by e.g. DVORAK et al., 1987, VOYIADJIS et al., 1995 and is discussed in SUQUET, 2014.

In the following, values referring to uniaxial tension, shear, uniaxial compression, equi-biaxial tension and equi-biaxial compression are labelled by the superscripts t, s, c, bt and bc.

A quadratic yield locus given by the SAMP model for the matrix material is shown in figure 5.3 alongside a von Mises definition in the $p - \sigma^{vM}$ plane. In case of the von Mises formulation, the yield locus for the isotropic matrix material m is determined by a line parallel to the p-axis through the input yield value for uniaxial tension $Y^{m,t}$. All the yield values are then predicted on the computed von Mises yield surface. When the SAMP model is applied, a quadratic yield locus is determined from the input yield values for uniaxial tension $Y^{m,t}$, shear $Y^{m,s}$ and uniaxial compression $Y^{m,c}$. Remaining yield values such as those for equi-biaxial tension $Y^{m,bt}$ or equi-biaxial compression $Y^{m,bc}$ can then be determined from the quadratic yield locus.

Applying expression (5.19), a multitude of the composite yield points can be defined for various load cases. Overall, these yield points define a yield surface for the pseudo grain composite under consideration. Such yield points are shown in the $\sigma_1 - \sigma_2$ yield locus in figure 5.4.

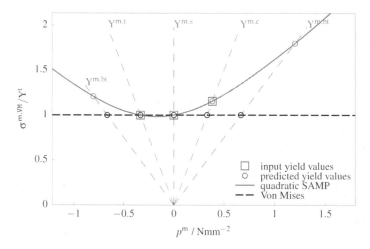

Figure 5.3 Yield surface definitions for the matrix material.

Since an incremental definition of the yield surface for all possible loading conditions would be infeasible for the further modelling process, an analytical expression for the yield surface is preferable. An analytical yield formulation could be derived from the tensor relations given previously. However, the application of a known yield surface formulation is preferable for the material model implementation since the required convexity constraints can be found in literature. When choosing an appropriate analytical formulation, the condition of general transferability to composites with multiple fibre volume fractions must be fulfilled. This includes the case of a composite with a negligible or non-existent amount of fibres. In this context, the analytical yield surface formulation of the composite should diminish to the original yield surface of the pure matrix from which it was derived.

In the context of this work, the SAMP approach given in expression (5.15) was chosen for the modelling of matrix yield. Therefore, it has to be recovered from the analytical anisotropic yield surface when $\vartheta^f = 0 \rightarrow \mathbb{T} = \boldsymbol{\delta}$. KOLLING et al., 2005 show that the SAMP yield formulation is derived as isotropic formulation of a general anisotropic quadratic yield surface model reminiscent of the Tsai-Wu formulation given by TSAI et al., 1971. Hence, the general anisotropic quadratic Tsai-Wu formulation is picked in order to describe the yield envelope of the composite analytically, seeing that this would fulfil the constraint mentioned for $\mathbb{T} = \boldsymbol{\delta}$.

The yield envelope of fibre reinforced composites is commonly conceived to be of orthotropic nature, as shown by VOGLER, 2014 among others. For an orthotropic material the Tsai-Wu yield surface is defined in stress space by equation (3.58). The terms $\vec{F}$ and $\boldsymbol{F}$ represent the Voigt notation variants of the respective second and fourth order strength tensors $\mathbf{F}$ and $\mathbb{F}$. Similarly, as shown by TSAI et al., 1971, the yield surface can be described in strain space by

$$\Phi = \vec{K}^{\mathrm{T}}\vec{\varepsilon} + \vec{\varepsilon}^{\mathrm{T}}\mathbf{K}\vec{\varepsilon} \quad = 1 \tag{5.20}$$

Applying the stiffness tensor C, the parameters $\vec{K}$ and $\mathbf{K}$ can be computed from $\vec{F}$ and $\mathbb{F}$ by

$$\vec{K} = C\vec{F} \quad \text{and} \quad \mathbf{K} = (\mathbf{C})^{\mathrm{T}}\mathbb{F}\mathbf{C}. \tag{5.21}$$

The variables $\vec{K}$ and $\mathbf{K}$ represent the Voigt's notation of the second respectively fourth order Tsai-Wu parameter tensors $\mathbf{K}$ and $\mathbb{K}$.

TSAI et al., 1971 list explicit equations for the calculation of the parameters for the orthotropic Tsai-Wu formulation from related engineering strengths. The expressions for tension Y_1^{t}, Y_2^{t}, Y_3^{t}, compression Y_1^{c}, Y_2^{c}, Y_3^{c}, shear Y_{12}^{s}, Y_{13}^{s}, Y_{23}^{s} and equi-biaxial tension Y_{12}^{bt}, Y_{13}^{bt}, Y_{23}^{bt} can be used for the computation of the required parameters as

$$
\begin{aligned}
F_1 &= 1/Y_1^{\mathrm{t}} - 1/Y_1^{\mathrm{c}}, & F_2 &= 1/Y_2^{\mathrm{t}} - 1/Y_2^{\mathrm{c}}, & F_3 &= 1/Y_3^{\mathrm{t}} - 1/Y_3^{\mathrm{c}} \\
F_{11} &= 1/\left(Y_1^{\mathrm{t}}Y_1^{\mathrm{c}}\right), & F_{22} &= 1/\left(Y_2^{\mathrm{t}}Y_2^{\mathrm{c}}\right), & F_{33} &= 1/\left(Y_3^{\mathrm{t}}Y_3^{\mathrm{c}}\right) \\
F_{44} &= 1/\left(Y_{23}^{\mathrm{s}}\right)^2, & F_{55} &= 1/\left(Y_{13}^{\mathrm{s}}\right)^2, & F_{66} &= 1/\left(Y_{12}^{\mathrm{s}}\right)^2
\end{aligned}
\tag{5.22}
$$

and

$$
\begin{aligned}
F_{12} &= 1/\left(2\left(Y_{12}^{\mathrm{bt}}\right)^2\right)\left(1 - Y_{12}^{\mathrm{bt}}(F_1 + F_2) - \left(Y_{12}^{\mathrm{bt}}\right)^2(F_{11} + F_{22})\right) \\
F_{13} &= 1/\left(2\left(Y_{13}^{\mathrm{bt}}\right)^2\right)\left(1 - Y_{13}^{\mathrm{bt}}(F_1 + F_3) - \left(Y_{13}^{\mathrm{bt}}\right)^2(F_{11} + F_{33})\right) \\
F_{23} &= 1/\left(2\left(Y_{23}^{\mathrm{bt}}\right)^2\right)\left(1 - Y_{23}^{\mathrm{bt}}(F_2 + F_3) - \left(Y_{23}^{\mathrm{bt}}\right)^2(F_{22} + F_{33})\right).
\end{aligned}
\tag{5.23}
$$

It is assumed that positive and negative shear strengths are equal in the principal axis system.

Since the pseudo grains consist of UD fibres, their respective yield envelope should be transversely isotropic. The Tsai-Wu formulation can be reduced to transverse isotropy by setting

$$F_3 = F_2, \ F_{13} = F_{12}, \ F_{33} = F_{22}, \ F_{55} = F_{66} \text{ and } F_{44} = 2(F_{22} - F_{23}). \tag{5.24}$$

Expressions (5.22) and (5.23) could be used to derive the parameters for the analytical yield surface formulation of the UD pseudo grain composite by application of its respective yield values for tension $Y_1^{\mathrm{UD,t}}$, $Y_2^{\mathrm{UD,t}}$, $Y_3^{\mathrm{UD,t}}$, compression $Y_1^{\mathrm{UD,c}}$, $Y_2^{\mathrm{UD,c}}$, $Y_3^{\mathrm{UD,c}}$, shear $Y_{12}^{\mathrm{UD,s}}$,

$Y_{13}^{UD,s}$, $Y_{23}^{UD,s}$ and equi-biaxial tension $Y_{12}^{UD,bt}$, $Y_{13}^{UD,bt}$, $Y_{23}^{UD,bt}$. These engineering strengths can be calculated from (5.19).

Depending on the definition of the matrix yield surface, the overall composite yield surface could have a non-convex form. This is for example the case for a non-convex matrix yield surface as can be defined by the quadratic or piecewise linear SAMP formulation given in KOLLING et al., 2005. In such a case, the UD pseudo grain yield surface that can be calculated by a multitude of composite yield points for various load cases by expression (5.19) will also have a non-convex form. In this context, the derivation of a global analytical yield surface from the few strength values given in (5.22) and (5.23) would result in an in-adequate approximation of the discrete micro-mechanical yield points that are computed by expression (5.19). To ensure a better approximation by the overall analytical yield formulation, a parameter-fitting approach is suggested. All of the multiple yield points of the discrete micro-mechanical yield surface determined by (5.19) can be taken into account by adjusting the parameters of the analytical Tsai-Wu yield surface of equation (3.58) to fit the incremental yield surface via a least squares method. Such a procedure has been implemented in the final material model.

In case of a convex quadratic SAMP definition by (5.15), the fitting procedure leads to the same Tsai-Wu parameters as can be calculated by (5.22) and (5.23) from the specific yield values found by (5.19).

From the previously defined steps, the parameters $\vec{F}^{UD}$ and $\boldsymbol{F}^{UD}$ of the analytical Tsai-Wu yield formulation for each UD pseudo grain are determined. By applying the stiffness tensor $\boldsymbol{C}^{UD}$ of the UD pseudo grain, the respective strain based form of the Tsai-Wu formulation can be computed by adjusting expression (5.21) to

$$\vec{K}^{UD} = \vec{F}^{UD} \boldsymbol{C} \quad \text{and} \quad \boldsymbol{K}^{UD} = \left(\boldsymbol{C}^{UD}\right)^{\mathrm{T}} \boldsymbol{F}^{UD} \boldsymbol{C}^{UD}. \tag{5.25}$$

The convexity of the analytical yield surface should be ensured in order to achieve determined finite yield values for all load cases. Furthermore, convexity of the yield surface assures numerical stability of the return mapping algorithm shown in section 5.5 that was implemented to establish the consistency condition. In TSAI et al., 1971 necessary constraints on the yield surface parameters for the assurance of convexity are given. TANG, 1979 notices that these are not sufficient to ensure a closed ellipsoidal yield surface and states that $\boldsymbol{F}$ must be positive and semi-definite to fulfil this prerequisite.

Instability of the yield surface can be checked by testing different boundary conditions of the parameters of the orthotropic yield surface. These conditions are derived in TANG, 1979. In the context of the implemented material model, stability of the yield surface is enforced by adjusting the yield surface parameters when an instability is detected. The instability conditions and the respective parameter adjustments are given in algorithm 5.1.

It can be seen that the off-axis interaction terms are modified to ensure convexity. In the numerical context, adjusted parameters are modified by a factor of 0.9 to increase the

Algorithm 5.1   Assurance of convexity of the yield surface.

---

1 **if** $F_{ii} \leq 0$ (with $i = 1,2,3$) $\rightarrow$ avoided by definition

2 **if** $F_{22}F_{33} - F_{23}^2 \leq 0$ **then** $F_{23} = sgn(F_{23})\sqrt{F_{22}F_{33}}$

3 **if** $F_{11}F_{33} - F_{13}^2 \leq 0$ **then** $F_{13} = sgn(F_{13})\sqrt{F_{11}F_{33}}$

4 **if** $F_{11}F_{22} - F_{12}^2 \leq 0$ **then** $F_{12} = sgn(F_{12})\sqrt{F_{22}F_{33}}$

5 **if** $F_{11}F_{22}F_{33} + 2F_{12}F_{23}F_{13} - F_{11}F_{23}^2 - F_{22}F_{13}^2 - F_{33}F_{12}^2 \leq 0$ **then**

    if $F_{23} \equiv max(F_{23}, F_{13}, F_{12})$ **then**

$$F_{23} = \frac{F_{12}F_{13}}{F_{11}} + sgn(F_{23})\left(\frac{F_{12}^2 F_{13}^2}{F_{11}^2} + F_{22}F_{33} - \frac{F_{22}F_{13}^2 - F_{33}F_{12}^2}{F_{11}}\right)^{1/2}$$

    else if $F_{13} \equiv max(F_{23}, F_{13}, F_{12})$ **then**

$$F_{13} = \frac{F_{12}F_{23}}{F_{22}} + sgn(F_{13})\left(\frac{F_{12}^2 F_{23}^2}{F_{22}^2} + F_{11}F_{33} - \frac{F_{11}F_{23}^2 - F_{33}F_{12}^2}{F_{22}}\right)^{1/2}$$

    else if $F_{12} \equiv max(F_{23}, F_{13}, F_{12})$ **then**

$$F_{12} = \frac{F_{13}F_{23}}{F_{33}} + sgn(F_{12})\left(\frac{F_{13}^2 F_{23}^2}{F_{33}^2} + F_{11}F_{22} - \frac{F_{11}F_{23}^2 - F_{22}F_{13}^2}{F_{33}}\right)^{1/2}$$

    end

end

---

stability of the yield surface and avoid numerical issues preventing convexity. Principally, this leads to an adjustment of biaxial strength values that fulfil the yield condition.

The overall yield surface computation process for each UD pseudo grain is visualised in figure 5.4, wherein the yield locus in the $\sigma_1 - \sigma_2$ plane is shown. Firstly, discrete micro-mechanical yield points were determined by incrementally applying equation (5.19) for different arbitrary stress states defined by (5.17). The entirety of these discrete yield points defines the micro-mechanical yield surface. The parameters of the analytical Tsai-Wu yield surface model are then adjusted to best fit the micro-mechanical yield surface. In this last step the convexity constraints of algorithm 5.1 are applied to the Tsai-Wu parameters. In case of a convex matrix yield surface and hence a convex micro-mechanical yield surface of the UD pseudo grain, the yield points for uniaxial tension and compression that can be extracted from the ellipsoidal Tsai-Wu yield surface correspond to the respective yield points of the micro-mechanical yield surface. This is shown for $Y_1^{UD,t}$, $Y_2^{UD,t}$ and compression $Y_1^{UD,c}$, $Y_2^{UD,c}$ in figure 5.4.

Once the analytical formulation of the pseudo grain yield envelope is defined, the second part of the two-step homogenisation process can be performed. Herein, the pseudo grain data has to be remapped to the original composite by means of orientation averaging. The orientation averaging is performed similarly to the stiffness tensor as shown in expression (5.13).

The strain based instead of the stress based Tsai-Wu formulation is used for the orientation averaging. This is due to the fact that Voigt's averaging, and hence the assumption of equal strain in all pseudo grains, was already stated to be preferable for the modelling

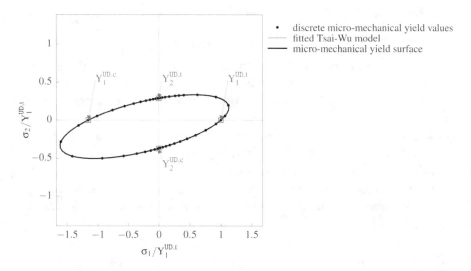

Figure 5.4 Tsai-Wu yield locus and corresponding micro-mechanical yield locus of the UD pseudo grain.

of the linear elastic response. Thus, the limit of purely elastic deformation should be assessed by an equal approach. This procedure essentially follows the strategy presented in VAN HATTUM et al., 1999 or LASPALAS et al., 2007. While the linear elastic model in LASPALAS et al., 2007 is based on the Tandon-Weng form of the Mori-Tanaka scheme, a differing approach was used by the authors for strength prediction. Here, LASPALAS et al., 2007 propose a rule of mixtures based model as was introduced in VAN HATTUM et al., 1999 for the pseudo grain calculations. The present work follows a more straight-forward approach by basing the micro-mechanical estimate of yield strength on the same micro-mechanical expressions applied for the computation of linear elastic behaviour.

Even though ADVANI et al., 1985 state that the orientation averaging procedure cannot precisely predict strength properties, it was perceived by VAN HATTUM et al., 1999 and LASPALAS et al., 2007 among others that it does give useful estimates for the determination of the boundary of linear elasticity.

The Tsai-Wu parameter tensors of the UD pseudo grain are of transversely isotropic nature. Thus, the orientation averaging for the fourth order tensor $\mathbf{K}$ can be performed in a similar fashion to the expressions given for the stiffness tensor in equation (5.14). Adopted from ADVANI et al., 1987, orientation averaging of the second order Tsai-Wu parameter tensor $\mathbf{K}$ is defined for each of its components as

$$K_{ij} = \left(K_{11}^{UD} - K_{22}^{UD} \right) A_{ij} + K_{22}^{UD} \delta_{ij}, \quad \text{with } i, j = 1, 2, 3. \tag{5.26}$$

The composite yield formulation can be recast to the stress based form by application of equation (5.21).

After the two-step homogenisation scheme is performed regarding the limit of purely linear elastic behaviour, respectively yield stress, an analytical Tsai-Wu yield surface formulation Φ is available for the oriented composite. The final analytical composite yield surface has to be constrained to convexity by applying the expressions given in algorithm 5.1.

Depending on the definition of the orientation tensor applied for orientation averaging in respect to the overall coordinate system, all components of the parameter tensors in Voigt's notation $\boldsymbol{F}$ and $\vec{F}$ may have non-zero values.

For the material model implementation, partial derivatives of the yield formulation Φ with respect to components of the stress tensor are necessary. These are given for normal stresses as

$$\frac{\partial \Phi}{\partial \sigma_1} = F_1 + 2(F_{11}\sigma_1 + F_{12}\sigma_2 + F_{13}\sigma_3) + \{2(F_{14}\sigma_4 + F_{15}\sigma_5 + F_{16}\sigma_6)\},$$

$$\frac{\partial \Phi}{\partial \sigma_2} = F_2 + 2(F_{22}\sigma_2 + F_{12}\sigma_1 + F_{23}\sigma_3) + \{2(F_{24}\sigma_4 + F_{25}\sigma_5 + F_{26}\sigma_6)\},$$

$$\frac{\partial \Phi}{\partial \sigma_3} = F_3 + 2(F_{33}\sigma_3 + F_{13}\sigma_1 + F_{23}\sigma_2) + \{2(F_{34}\sigma_4 + F_{35}\sigma_5 + F_{36}\sigma_6)\}$$

$$(5.27)$$

and for shear stresses as

$$\frac{\partial \Phi}{\partial \sigma_4} = 2F_{44}\sigma_4 + \{2(F_{14}\sigma_4 + F_{24}\sigma_4 + F_{34}\sigma_4 + F_{45}\sigma_4 + F_{56}\sigma_6) + F_4\},$$

$$\frac{\partial \Phi}{\partial \sigma_5} = 2F_{55}\sigma_5 + \{2(F_{15}\sigma_5 + F_{25}\sigma_5 + F_{35}\sigma_5 + F_{45}\sigma_5 + F_{56}\sigma_5) + F_5\}, \quad (5.28)$$

$$\frac{\partial \Phi}{\partial \sigma_6} = 2F_{66}\sigma_6 + \{2(F_{16}\sigma_6 + F_{26}\sigma_6 + F_{36}\sigma_6 + F_{46}\sigma_6 + F_{56}\sigma_6) + F_6\}.$$

In case the coordinate system of examination coincides with the principal axis of the orthotropic material system, the terms in curly brackets will result to zero. In the following, it is assumed that the composite volume is always examined in the principal material axis system. This system coincides with the principal axis system of the orientation tensor.

5.3.2 Flow Potential

Non-associated flow is considered for the material model, cf. section 3.4. As is shown in e.g. HOFFARTH, 2016, associated flow could lead to non-physical definitions of plastic Poisson's ratio. The term plastic Poisson's ratio refers to the relation of plastic strains as defined in section 3.4. A further discussion of the inelastic Poisson's ratio can be found e.g. in KOLUPAEV, 2018.

The flow potential formulation is defined similarly to the yield surface. However, the pressure dependent terms are omitted and the square root of the polynomial equation is defined as the flow potential. An orthotropic flow potential h with the parameter tensor $\boldsymbol{H}$ in Voigt's notation is assumed for the overall composite. This leads to

$$
\begin{aligned}
h = \sqrt{\vec{\sigma}^{\mathsf{T}}\boldsymbol{H}\vec{\sigma}} \\
= \big(H_{11}\sigma_1^2 + H_{22}\sigma_2^2 + H_{33}\sigma_3^2 + 2H_{12}\sigma_1\sigma_2 \\
+ 2H_{13}\sigma_1\sigma_3 + 2H_{23}\sigma_2\sigma_3 + H_{44}\sigma_4^2 + H_{55}\sigma_5^2 + H_{66}\sigma_6^2\big)^{1/2}.
\end{aligned}
\tag{5.29}
$$

The description in equation (5.29) is reminiscent of the Hill yield surface (HILL, 1948).

The omission of the pressure dependent terms results in the plastic Poisson's ratios being independent of stress, as will be shown at the end of this section. This allows a simplified material model derivation and avoids numerical issues, cf. HOFFARTH et al., 2017.

Convexity of the flow potential has to be ensured to assure the validity of the material model in respect to the second law of thermodynamics, cf. BETTEN, 1979. This is established by following the procedure defined for the yield surface in algorithm 5.1.

Partial derivatives of the flow potential with respect to components of the stress tensor are given for normal stresses as

$$
\begin{aligned}
\frac{\partial h}{\partial \sigma_1} &= \frac{H_{11}\sigma_1 + H_{12}\sigma_2 + H_{13}\sigma_3}{h}, \\
\frac{\partial h}{\partial \sigma_2} &= \frac{H_{22}\sigma_2 + H_{12}\sigma_1 + H_{23}\sigma_3}{h}, \\
\frac{\partial h}{\partial \sigma_3} &= \frac{H_{33}\sigma_3 + H_{13}\sigma_1 + H_{23}\sigma_2}{h}
\end{aligned}
\tag{5.30}
$$

and for shear stresses as

$$
\frac{\partial h}{\partial \sigma_4} = \frac{H_{44}\sigma_4}{h}, \quad \frac{\partial h}{\partial \sigma_5} = \frac{H_{55}\sigma_5}{h} \quad \text{and} \quad \frac{\partial h}{\partial \sigma_6} = \frac{H_{66}\sigma_6}{h}.
\tag{5.31}
$$

It can be seen that the derivative of the flow potential with respect to stresses can be simplified to the form

$$
\frac{\partial h}{\partial \vec{\sigma}} = \frac{\boldsymbol{H}\vec{\sigma}}{h}.
\tag{5.32}
$$

The Euclidean norm of the flow potential derivative is defined as

$$
\begin{aligned}
\left\| \frac{\partial h}{\partial \vec{\sigma}} \right\| &= \left(\left(\frac{\partial h}{\partial \sigma_1} \right)^2 + \left(\frac{\partial h}{\partial \sigma_2} \right)^2 + \left(\frac{\partial h}{\partial \sigma_3} \right)^2 \right. \\
&\qquad \left. + \left(\frac{\partial h}{\partial \sigma_4} \right)^2 + \left(\frac{\partial h}{\partial \sigma_5} \right)^2 + \left(\frac{\partial h}{\partial \sigma_6} \right)^2 \right)^{1/2} \\
&= \frac{1}{h} \left((H_{11}\sigma_1 + H_{12}\sigma_2 + H_{13}\sigma_3)^2 \right. \\
&\qquad + (H_{22}\sigma_2 + H_{12}\sigma_1 + H_{23}\sigma_3)^2 \\
&\qquad + (H_{33}\sigma_3 + H_{13}\sigma_1 + H_{23}\sigma_2)^2 \\
&\qquad \left. + (H_{44}\sigma_4)^2 + (H_{55}\sigma_5)^2 + (H_{66}\sigma_6)^2 \right)^{1/2} \\
&= \frac{\sqrt{(\boldsymbol{H}\vec{\sigma})^{\mathrm{T}}(\boldsymbol{H}\vec{\sigma})}}{h}.
\end{aligned}
\tag{5.33}
$$

In the following, the normalised flow potential derivative values are derived for different stress states.

In case of in-plane uniaxial loading with $\sigma_1 \neq 0$ and $\sigma_{i\neq1} = 0$, the normalised derivative results in

$$
\begin{aligned}
\frac{\partial h}{\partial \sigma_1} \left\| \frac{\partial h}{\partial \vec{\sigma}} \right\|^{-1} &= \frac{H_{11}\sigma_1}{\left(H_{11}^2 + H_{12}^2 + H_{13}^2 \right)^{1/2} \sigma_1}, \\
\frac{\partial h}{\partial \sigma_2} \left\| \frac{\partial h}{\partial \vec{\sigma}} \right\|^{-1} &= \frac{H_{12}\sigma_1}{\left(H_{11}^2 + H_{12}^2 + H_{13}^2 \right)^{1/2} \sigma_1}, \\
\frac{\partial h}{\partial \sigma_3} \left\| \frac{\partial h}{\partial \vec{\sigma}} \right\|^{-1} &= \frac{H_{13}\sigma_1}{\left(H_{11}^2 + H_{12}^2 + H_{13}^2 \right)^{1/2} \sigma_1}.
\end{aligned}
\tag{5.34}
$$

With $\sigma_2 \neq 0$ and $\sigma_{i\neq2} = 0$, it is determined as

$$
\begin{aligned}
\frac{\partial h}{\partial \sigma_1} \left\| \frac{\partial h}{\partial \vec{\sigma}} \right\|^{-1} &= \frac{H_{12}\sigma_2}{\left(H_{22}^2 + H_{12}^2 + H_{23}^2 \right)^{1/2} \sigma_2}, \\
\frac{\partial h}{\partial \sigma_2} \left\| \frac{\partial h}{\partial \vec{\sigma}} \right\|^{-1} &= \frac{H_{22}\sigma_2}{\left(H_{22}^2 + H_{12}^2 + H_{23}^2 \right)^{1/2} \sigma_2}, \\
\frac{\partial h}{\partial \sigma_2} \left\| \frac{\partial h}{\partial \vec{\sigma}} \right\|^{-1} &= \frac{H_{23}\sigma_2}{\left(H_{22}^2 + H_{12}^2 + H_{23}^2 \right)^{1/2} \sigma_2}.
\end{aligned}
\tag{5.35}
$$

When uniaxial loading is given by $\sigma_3 \neq 0$ and $\sigma_{i\neq3} = 0$, the derivative is calculated as

$$\frac{\partial h}{\partial \sigma_1}\left\|\frac{\partial h}{\partial \vec{\sigma}}\right\|^{-1} = \frac{H_{13}\sigma_3}{\left(H_{33}^2 + H_{13}^2 + H_{23}^2\right)^{1/2}\sigma_3},$$

$$\frac{\partial h}{\partial \sigma_2}\left\|\frac{\partial h}{\partial \vec{\sigma}}\right\|^{-1} = \frac{H_{23}\sigma_3}{\left(H_{33}^2 + H_{13}^2 + H_{23}^2\right)^{1/2}\sigma_3}, \qquad (5.36)$$

$$\frac{\partial h}{\partial \sigma_3}\left\|\frac{\partial h}{\partial \vec{\sigma}}\right\|^{-1} = \frac{H_{33}\sigma_3}{\left(H_{33}^2 + H_{13}^2 + H_{23}^2\right)^{1/2}\sigma_3}.$$

The normalised derivative for shear loading with $\sigma_4 \neq 0$, $\sigma_{i\neq4} = 0$ amounts to

$$\frac{\partial h}{\partial \sigma_4}\left\|\frac{\partial h}{\partial \vec{\sigma}}\right\|^{-1} = \frac{H_{44}\sigma_4}{\left(H_{44}^2\right)^{1/2}\sigma_4}. \qquad (5.37)$$

In case shear loading is defined by $\sigma_5 \neq 0$ and $\sigma_{i\neq5} = 0$, the derivative is given as

$$\frac{\partial h}{\partial \sigma_5}\left\|\frac{\partial h}{\partial \vec{\sigma}}\right\|^{-1} = \frac{H_{55}\sigma_5}{\left(H_{55}^2\right)^{1/2}\sigma_5}. \qquad (5.38)$$

For $\sigma_6 \neq 0$ and $\sigma_{i\neq6} = 0$ it can be calculated by

$$\frac{\partial h}{\partial \sigma_6}\left\|\frac{\partial h}{\partial \vec{\sigma}}\right\|^{-1} = \frac{H_{66}\sigma_6}{\left(H_{66}^2\right)^{1/2}\sigma_6}. \qquad (5.39)$$

Equi-biaxial loading with $\sigma_1 = \sigma_2$ and $\sigma_{i\neq1,2} = 0$ resolves to

$$\frac{\partial h}{\partial \sigma_1}\left\|\frac{\partial h}{\partial \vec{\sigma}}\right\|^{-1} = \frac{(H_{11}+H_{12})\sigma_1}{\left((H_{11}+H_{12})^2 + (H_{22}+H_{12})^2 + (H_{13}+H_{23})^2\right)^{1/2}\sigma_1},$$

$$\frac{\partial h}{\partial \sigma_2}\left\|\frac{\partial h}{\partial \vec{\sigma}}\right\|^{-1} = \frac{(H_{22}+H_{12})\sigma_1}{\left((H_{11}+H_{12})^2 + (H_{22}+H_{12})^2 + (H_{13}+H_{23})^2\right)^{1/2}\sigma_1}. \qquad (5.40)$$

It can be seen in equations (5.34) - (5.40) that the derivative values are independent of stress. This has a considerable advantage for the derivation of the material model. The normalisation leads to simple relations for the shear derivatives. As described in equation (3.52), the increment of plastic strain is determined by the flow rule. Application of the normalised flow potential derivative and Voigt's notation to expression (3.52) leads to

$$d\vec{\varepsilon}^{\text{pl}} = d\lambda \frac{\partial h}{\partial \vec{\sigma}} = d\bar{\lambda} \frac{\partial h}{\partial \vec{\sigma}}\left\|\frac{\partial h}{\partial \vec{\sigma}}\right\|^{-1}. \qquad (5.41)$$

Applying expression (3.53) given in section 3.4 and using the definition of equation (5.41), the plastic Poisson's ratios v_{ij}^{pl} (with $i, j = 1, 2, 3$) can be derived for uniaxial loading as

$$
\begin{aligned}
v_{12}^{\text{pl}} &= -\frac{d\varepsilon_2^{\text{pl}}}{d\varepsilon_1^{\text{pl}}} = -\frac{H_{12}}{H_{11}}, & v_{13}^{\text{pl}} &= -\frac{d\varepsilon_3^{\text{pl}}}{d\varepsilon_1^{\text{pl}}} = -\frac{H_{13}}{H_{11}}, \\
v_{21}^{\text{pl}} &= -\frac{d\varepsilon_1^{\text{pl}}}{d\varepsilon_2^{\text{pl}}} = -\frac{H_{12}}{H_{22}}, & v_{23}^{\text{pl}} &= -\frac{d\varepsilon_3^{\text{pl}}}{d\varepsilon_2^{\text{pl}}} = -\frac{H_{23}}{H_{22}}, \\
v_{31}^{\text{pl}} &= -\frac{d\varepsilon_1^{\text{pl}}}{d\varepsilon_3^{\text{pl}}} = -\frac{H_{13}}{H_{33}}, & v_{32}^{\text{pl}} &= -\frac{d\varepsilon_2^{\text{pl}}}{d\varepsilon_3^{\text{pl}}} = -\frac{H_{23}}{H_{33}}.
\end{aligned}
\tag{5.42}
$$

The evaluation of the flow parameters H is presented in section 5.3.7.

5.3.3 Effective Plastic Strain

Effective strains are derived by use of the equivalence of work, BERG, 1972. In the case of incremental plastic loading, plastic work equivalence is defined so that the incremental work performed has to be equal to the product of effective stress $\overline{\sigma}$ and incremental effective plastic strain $d\overline{\varepsilon}^{\text{pl}}$ as

$$
\vec{\sigma}^{\text{T}} d\vec{\varepsilon}^{\text{pl}} = \overline{\sigma}\, d\overline{\varepsilon}^{\text{pl}}.
\tag{5.43}
$$

By using the results of equations (5.32) and (5.33), the flow rule definition of equation (5.41) can be written as

$$
d\vec{\varepsilon}^{\text{pl}} = d\overline{\lambda}\frac{\boldsymbol{H}\vec{\sigma}}{h}\left(\frac{\sqrt{(\boldsymbol{H}\vec{\sigma})^{\text{T}}(\boldsymbol{H}\vec{\sigma})}}{h}\right)^{-1} = d\overline{\lambda}\frac{\boldsymbol{H}\vec{\sigma}}{\sqrt{(\boldsymbol{H}\vec{\sigma})^{\text{T}}(\boldsymbol{H}\vec{\sigma})}}.
\tag{5.44}
$$

Taking into account the results from equations (5.29) and (5.44) and applying these to the equivalence of plastic work (5.43), the plastic multiplier can be related to the incremental effective plastic strain as

$$
\begin{aligned}
d\overline{\varepsilon}^{\text{pl}} &= \frac{\vec{\sigma}^{\text{T}} d\vec{\varepsilon}^{\text{pl}}}{\overline{\sigma}} = d\overline{\lambda}\frac{\vec{\sigma}^{\text{T}}(\boldsymbol{H}\vec{\sigma})}{\overline{\sigma}\sqrt{(\boldsymbol{H}\vec{\sigma})^{\text{T}}(\boldsymbol{H}\vec{\sigma})}} = d\overline{\lambda}\frac{h^2}{\overline{\sigma}\sqrt{(\boldsymbol{H}\vec{\sigma})^{\text{T}}(\boldsymbol{H}\vec{\sigma})}} \\
&\rightarrow d\overline{\lambda} = d\overline{\varepsilon}^{\text{pl}}\frac{\overline{\sigma}\sqrt{(\boldsymbol{H}\vec{\sigma})^{\text{T}}(\boldsymbol{H}\vec{\sigma})}}{h^2}.
\end{aligned}
\tag{5.45}
$$

The expression for the plastic multiplier given in (5.45) can be included into the flow rule (5.44) to derive the relation between stress and plastic strain increment as

$$d\vec{\varepsilon}^{\mathrm{pl}} = d\bar{\varepsilon}^{\mathrm{pl}} \frac{\bar{\sigma} \boldsymbol{H} \vec{\sigma}}{h^2}$$

$$\rightarrow \vec{\sigma} = \frac{h^2}{\bar{\sigma} d\bar{\varepsilon}^{\mathrm{pl}}} \boldsymbol{H}^{-1} d\vec{\varepsilon}^{\mathrm{pl}}. \tag{5.46}$$

By substituting the stress vector in the equivalence of plastic work (5.43) with relation (5.46) and considering the symmetry of $\boldsymbol{H}$, the incremental effective plastic strain is derived as

$$d\bar{\varepsilon}^{\mathrm{pl}} = \frac{\vec{\sigma}^{\mathrm{T}} d\vec{\varepsilon}^{\mathrm{pl}}}{\bar{\sigma}} = \frac{h^2}{\bar{\sigma}^2 d\bar{\varepsilon}^{\mathrm{pl}}} (\boldsymbol{H}^{-1} d\vec{\varepsilon}^{\mathrm{pl}})^{\mathrm{T}} d\vec{\varepsilon}^{\mathrm{pl}}$$

$$= \frac{h}{\bar{\sigma}} \left((d\vec{\varepsilon}^{\mathrm{pl}})^{\mathrm{T}} \boldsymbol{H}^{-1} d\vec{\varepsilon}^{\mathrm{pl}} \right)^{1/2}. \tag{5.47}$$

5.3.4 Effective Stress

The effective stress $\bar{\sigma}$ is defined to be equal to the stress determined in longitudinal direction under uniaxial tension Y_1^{t}:

$$\bar{\sigma} = Y_1^{\mathrm{t}}. \tag{5.48}$$

Furthermore, effective stress is defined to be equal to the flow potential h which was given in equation (5.29). This is expressed by

$$\bar{\sigma} = h. \tag{5.49}$$

Hence, the expression for effective plastic strain in (5.47) can be simplified.

Recently, HOFFARTH et al., 2017 proposed a similar definition of yield surface and flow potential for orthotropic materials with an equivalent formulation for the effective stress.

5.3.5 Relation of Plastic Strains and Effective Plastic Strain

Expression (5.49) for the effective stress $\bar{\sigma} = h$ and expression (5.46) for the plastic strain increment can be combined to

$$
d\vec{\varepsilon}^{\,\mathrm{pl}} = d\bar{\varepsilon}^{\,\mathrm{pl}}\frac{H\vec{\sigma}}{h} =
\begin{bmatrix}
H_{11}\sigma_1 + H_{12}\sigma_2 + H_{13}\sigma_3 \\
H_{12}\sigma_1 + H_{22}\sigma_2 + H_{23}\sigma_3 \\
H_{13}\sigma_1 + H_{23}\sigma_2 + H_{33}\sigma_3 \\
H_{44}\sigma_4 \\
H_{55}\sigma_5 \\
H_{66}\sigma_6
\end{bmatrix}
\frac{d\bar{\varepsilon}^{\,\mathrm{pl}}}{h}.
\tag{5.50}
$$

E.g. for the uniaxial tensile loading case $\sigma_1 \neq 0$ and $\sigma_{i\neq1} = 0$ expression (5.50) resolves to

$$
d\vec{\varepsilon}^{\,\mathrm{pl}} = \left[H_{11}, H_{12}, H_{13}, 0, 0, 0\right]^{\mathrm{T}} \frac{d\bar{\varepsilon}^{\,\mathrm{pl}}}{\sqrt{H_{11}}}.
\tag{5.51}
$$

For shear loading with e.g. $\sigma_4 \neq 0$, $\sigma_{i\neq4} = 0$ it yields

$$
d\vec{\varepsilon}^{\,\mathrm{pl}} = \left[0, 0, 0, H_{44}, 0, 0\right]^{\mathrm{T}} \frac{d\bar{\varepsilon}^{\,\mathrm{pl}}}{\sqrt{H_{44}}}
\tag{5.52}
$$

and for equi-biaxial loading with e.g. $\sigma_1 = \sigma_2$ and $\sigma_{i\neq1,2} = 0$ it leads to

$$
d\vec{\varepsilon}^{\,\mathrm{pl}} = \left[H_{11} + H_{12}, H_{12} + H_{22}, H_{13} + H_{23}, 0, 0, 0\right]^{\mathrm{T}} \frac{d\bar{\varepsilon}^{\,\mathrm{pl}}}{\sqrt{H_{11} + H_{22} + 2H_{12}}}.
\tag{5.53}
$$

With these relations, the effective plastic strain increment can be related to the plastic strain increment acquired in the experiments. This is necessary to formulate the hardening relations proposed in section 5.3.6 in respect to effective plastic strain within the material model.

5.3.6 Hardening

Hardening is included by an adjustment of engineering strength values Y of the final yield surface.

Considering specific stress states $\vec{\sigma}$ in expression (3.58), the respective strength values Y of the Tsai-Wu yield surface can be calculated by considering

$$
\begin{aligned}
&\text{if } d \neq 0 \rightarrow Y = c/(2d) \pm \sqrt{c^2/(4d^2) + 1/d} \\
&\text{else} \rightarrow Y = 1/c, \quad \text{with } c = \vec{F}^\mathsf{T} \text{ and } d = (\vec{\sigma})^\mathsf{T} \boldsymbol{F} \vec{\sigma}.
\end{aligned}
\tag{5.54}
$$

Alternatively, the strengths for predefined stress states can be extracted from the Tsai-Wu parameters specified in expressions (5.22) and (5.23).

In order to consider hardening, the initial yield values Y_0 are modified by a predefined hardening relation. ψ designates an internal variable that controls hardening, respectively strengths Y that differ from the initial state, by

$$
Y = f(Y_0, \psi).
\tag{5.55}
$$

Analytical expressions for this relation are discussed in section 5.3.8.

The strength values for the hardened state are applied to compute a hardened yield surface with equation (3.58) by use of expressions (5.22) and (5.23). The stability conditions of algorithm 5.1 are enforced in every hardening step to ensure convexity of the yield surface in all states.

The hardening formulation is isotropic in the sense that an increase of strength for one stress state will lead to a strength increase for all stress states. This is due to the yield surface being scaled in its entirety by the internal variable. Yet, the amount of scaling for the different stress states may differ.

By applying the previous expressions, the yield formulation Φ depends on an internal variable in addition to the stress state. The derivative of the yield surface in respect to the internal variable is given as

$$
\begin{aligned}
\frac{\partial \Phi}{\partial \psi} = {} & \left(\frac{\partial Y_1^c / \partial \psi}{(Y_1^c)^2} - \frac{\partial Y_1^t / \partial \psi}{(Y_1^t)^2} \right) \sigma_1 \\
& + \left(\frac{\partial Y_2^c / \partial \psi}{(Y_2^c)^2} - \frac{\partial Y_2^t / \partial \psi}{(Y_2^t)^2} \right) \sigma_2 + \left(\frac{\partial Y_3^c / \partial \psi}{(Y_3^c)^2} - \frac{\partial Y_3^t / \partial \psi}{(Y_3^t)^2} \right) \sigma_3 \\
& - \left(\frac{\partial Y_1^t / \partial \psi}{Y_1^c (Y_1^t)^2} + \frac{\partial Y_1^c / \partial \psi}{Y_1^t (Y_1^c)^2} \right) \sigma_1^2 \\
& - \left(\frac{\partial Y_2^t / \partial \psi}{Y_2^c (Y_2^t)^2} + \frac{\partial Y_2^c / \partial \psi}{Y_2^t (Y_2^c)^2} \right) \sigma_2^2 - \left(\frac{\partial Y_3^t / \partial \psi}{Y_3^c (Y_3^t)^2} + \frac{\partial Y_3^c / \partial \psi}{Y_3^t (Y_3^c)^2} \right) \sigma_3^2 \\
& - \left(2 \frac{\partial Y_{23}^s / \partial \psi}{(Y_{23}^s)^3} \right) \sigma_4^2 - \left(2 \frac{\partial Y_{13}^s / \partial \psi}{(Y_{13}^s)^3} \right) \sigma_5^2 - \left(2 \frac{\partial Y_{12}^s / \partial \psi}{(Y_{12}^s)^3} \right) \sigma_6^2 \\
& + \left(-2 \frac{\partial Y_{12}^{bt} / \partial \psi}{(Y_{12}^{bt})^3} + \frac{\partial Y_{12}^{bt} / \partial \psi}{(Y_{12}^{bt})^2 Y_1^t} + \frac{\partial Y_1^t / \partial \psi}{(Y_1^t)^2 Y_{12}^{bt}} - \frac{\partial Y_{12}^{bt} / \partial \psi}{(Y_{12}^{bt})^2 Y_1^c} \right. \\
& \quad + \frac{\partial Y_1^c / \partial \psi}{(Y_1^c)^2 Y_{12}^{bt}} + \frac{\partial Y_{12}^{bt} / \partial \psi}{(Y_{12}^{bt})^2 Y_2^t} + \frac{\partial Y_2^t / \partial \psi}{(Y_2^t)^2 Y_{12}^{bt}} - \frac{\partial Y_{12}^{bt} / \partial \psi}{(Y_{12}^{bt})^2 Y_2^c} + \frac{\partial Y_2^c / \partial \psi}{(Y_2^c)^2 Y_{12}^{bt}} \\
& \quad \left. + \frac{\partial Y_1^t / \partial \psi}{(Y_1^t)^2 Y_1^c} + \frac{\partial Y_1^c / \partial \psi}{(Y_1^c)^2 Y_1^t} + \frac{\partial Y_2^t / \partial \psi}{(Y_2^t)^2 Y_2^c} + \frac{\partial Y_2^c / \partial \psi}{(Y_2^c)^2 Y_2^t} \right) \sigma_1 \sigma_2 \\
& + \left(-2 \frac{\partial Y_{13}^{bt} / \partial \psi}{(Y_{13}^{bt})^3} + \frac{\partial Y_{13}^{bt} / \partial \psi}{(Y_{13}^{bt})^2 Y_1^t} + \frac{\partial Y_1^t / \partial \psi}{(Y_1^t)^2 Y_{13}^{bt}} - \frac{\partial Y_{13}^{bt} / \partial \psi}{(Y_{13}^{bt})^2 Y_1^c} \right. \\
& \quad + \frac{\partial Y_1^c / \partial \psi}{(Y_1^c)^2 Y_{13}^{bt}} + \frac{\partial Y_{13}^{bt} / \partial \psi}{(Y_{13}^{bt})^2 Y_3^t} + \frac{\partial Y_3^t / \partial \psi}{(Y_3^t)^2 Y_{13}^{bt}} - \frac{\partial Y_{13}^{bt} / \partial \psi}{(Y_{13}^{bt})^2 Y_3^c} + \frac{\partial Y_3^c / \partial \psi}{(Y_3^c)^2 Y_{13}^{bt}} \\
& \quad \left. + \frac{\partial Y_1^t / \partial \psi}{(Y_1^t)^2 Y_1^c} + \frac{\partial Y_1^c / \partial \psi}{(Y_1^c)^2 Y_1^t} + \frac{\partial Y_3^t / \partial \psi}{(Y_3^t)^2 Y_3^c} + \frac{\partial Y_3^c / \partial \psi}{(Y_3^c)^2 Y_3^t} \right) \sigma_1 \sigma_3 \\
& + \left(-2 \frac{\partial Y_{23}^{bt} / \partial \psi}{(Y_{23}^{bt})^3} + \frac{\partial Y_{23}^{bt} / \partial \psi}{(Y_{23}^{bt})^2 Y_2^t} + \frac{\partial Y_2^t / \partial \psi}{(Y_2^t)^2 Y_{23}^{bt}} - \frac{\partial Y_{23}^{bt} / \partial \psi}{(Y_{23}^{bt})^2 Y_2^c} \right. \\
& \quad + \frac{\partial Y_2^c / \partial \psi}{(Y_2^c)^2 Y_{23}^{bt}} + \frac{\partial Y_{23}^{bt} / \partial \psi}{(Y_{23}^{bt})^2 Y_3^t} + \frac{\partial Y_3^t / \partial \psi}{(Y_3^t)^2 Y_{23}^{bt}} - \frac{\partial Y_{23}^{bt} / \partial \psi}{(Y_{23}^{bt})^2 Y_3^c} + \frac{\partial Y_3^c / \partial \psi}{(Y_3^c)^2 Y_{23}^{bt}} \\
& \quad \left. + \frac{\partial Y_2^t / \partial \psi}{(Y_2^t)^2 Y_2^c} + \frac{\partial Y_2^c / \partial \psi}{(Y_2^c)^2 Y_2^t} + \frac{\partial Y_2^t / \partial \psi}{(Y_2^t)^2 Y_3^c} + \frac{\partial Y_3^c / \partial \psi}{(Y_3^c)^2 Y_3^t} \right) \sigma_2 \sigma_3 .
\end{aligned}
\tag{5.56}
$$

The effective plastic strain derived in section 5.3.3 is specified as internal variable governing hardening for the material model.

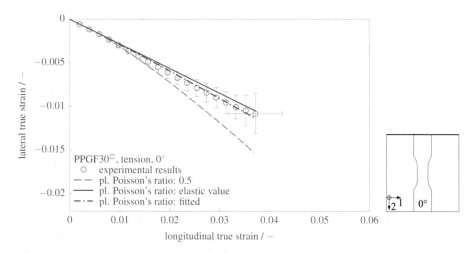

Figure 5.5   Relation of total lateral to longitudinal strain in 0° PPGF30$^\square$ samples.

5.3.7 Determination of Flow Parameters

The flow parameters can be determined by evaluating plastic Poisson's ratios. The evaluation of the elastic Poisson's ratio in the experiments was presented in section 4.3.2.4. Similarly, the plastic Poisson's ratio can be analysed by determining the slope of the curve of transverse strain plotted with respect to the longitudinal strain in the plastic region, cf. equation (5.42).

The incremental relation of plastic lateral strains to plastic longitudinal strains in the experiments actually varies throughout the deformation process. Nevertheless, in the numerical modelling context, a constant plastic Poisson's ratio is considered during deformation. This leads to constant parameters of the flow function for the entire deformation range.

Figure 5.5 shows the relation of total lateral strains to total longitudinal strains for 0° tensile loading of the PPGF30$^\square$ parts. A plastic Poisson's ratio can be determined from the experiments by approximating a constant slope that best fits the relation of plastic lateral to plastic longitudinal strains by application of a least squares method. The line labelled "pl. Poisson's ratio: fitted" was determined by utilising the thereby computed value. Over the entire deformation range it gives an adequate prediction of the experimental values.

Two additional lines are shown in the figure. The line labelled "pl. Poisson's ratio: 0.5" is the result of assuming an isochoric relation of plastic lateral and longitudinal strain. It is clear from the figure that such an assumption leads to large deviations in comparison to the experimental results. The curve "pl. Poisson's ratio: elastic value" was determined by assuming that the plastic Poisson's ratio equals the elastic Poisson's ratio. Figure 5.5 shows that this assumption leads to a deviation of 2.97 % of the predicted minimum total lateral strain in relation to the experimental value. The predicted strain remains in the region of the double standard deviation found in the experiments.

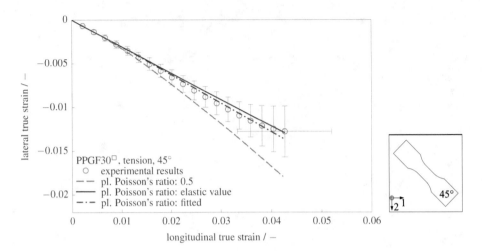

Figure 5.6   Relation of total lateral to longitudinal strain in 45° PPGF30□ samples.

Hence, it is assumed that the plastic Poisson's ratio can be approximately predicted from the elastic Poisson's ratio. This presumption decreases the effort necessary for material parametrisation.

The relations of total lateral to longitudinal strains for tensile samples of PPGF30□ loaded in 45° and 90° are plotted in figures 5.6 and 5.7. Results are found which are similar to the ones discussed for 0° loading. The assumption that the plastic Poisson's ratio can be approximated by the elastic Poisson's ratio leads to deviations of 1.1 %, respectively 8.2 % of the predicted minimal lateral strains in comparison to the experimental values.

For the highly oriented PPGF30◊ sample the total strain relation is shown in figure 5.8 for 0° loading and in figure 5.9 for 90° loading. In 0° the plastic Poisson's ratio can be adequately approximated by the elastic Poisson's ratio. The difference of 0.17 % between the predicted minimum lateral strain and the experimental value is negligible.

The transversely loaded flat bar sample in figure 5.9 shows the highest tensile deformation capability of the PPGF30 parts. In this context, the assumption of equal Poisson's ratios in the elastic and the plastic range gives only a very rough approximation. Up to total longitudinal strain values of 0.025 the predicted lateral strains lay in the region of the double standard deviations of the experiments. However, the prediction strongly deviates at increased strain levels. At the point of failure the predicted minimum lateral strain values deviate from the experiments by 38 %.

Finally, the total strain relations are also evaluated for the PP matrix material in figure 5.10. It is seen that the ratio varies throughout the entire deformation process. The approximation of the Poisson's ratio by a fitted constant value leads to an adequate representation at longitudinal strains >0.9. Nevertheless, up to this value the lateral strain values

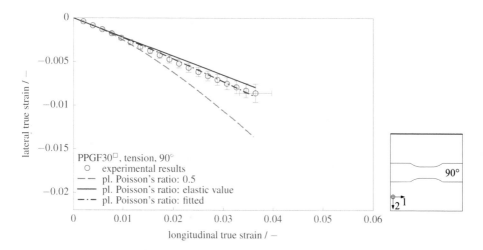

Figure 5.7   Relation of total lateral to longitudinal strain in 90° PPGF30$^\square$ samples.

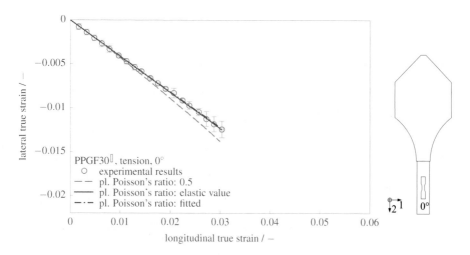

Figure 5.8   Relation of total lateral to longitudinal strain in 0° PPGF30$^\|$ samples.

are more adequately predicted by applying the elastic Poisson's ratio. It is clear that for higher deformations the Poisson's ratio should by considered as a variable depending on lateral strain in the modelling process.

In summary, for the PPGF30 material the plastic strain relations can be represented by a constant plastic Poisson's ratio value. The assumption of isochoric plastic deformation gives inadequate predictions for all evaluated parts. When fitted plastic Poisson's ratio values are not available, it can be assumed that plastic and elastic Poisson's ratios are equal.

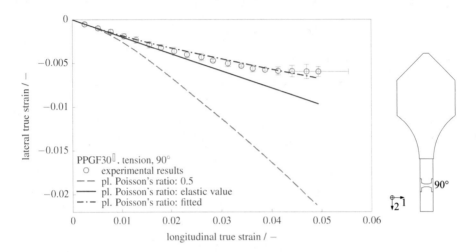

Figure 5.9   Relation of total lateral to longitudinal strain in 90° PPGF30▯ samples.

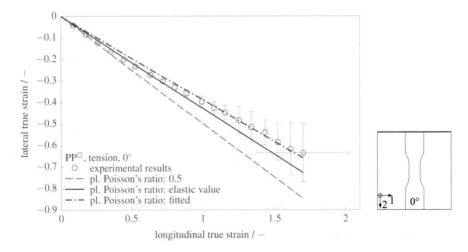

Figure 5.10   Relation of total lateral to longitudinal strain in 0° PP samples.

For most of the experiments this represents a reasonable approximation, especially in the case of small deformations. Taking into account the conclusions derived in section 4.3.2.3 and considering that the overall plastic deformation in PPGF30 parts is small, this approach was applied in the proposed material model.

Applying the previous assumptions to expression (5.42), the parameters of the flow surface can be derived as

$$H_{12} = -v_{12}^{pl} H_{11} \approx -v_{12}^{el} H_{11},$$

$$H_{13} = -v_{13}^{pl} H_{11} \approx -v_{13}^{el} H_{11},$$

$$H_{23} = -v_{23}^{pl} \frac{v_{12}^{pl}}{v_{21}^{pl}} H_{11} \approx -v_{23}^{el} \frac{v_{12}^{el}}{v_{21}^{el}} H_{11},$$

$$H_{22} = \frac{v_{12}^{pl}}{v_{21}^{pl}} H_{11} \approx \frac{v_{12}^{el}}{v_{21}^{el}} H_{11},$$

$$H_{33} = \frac{v_{13}^{pl}}{v_{31}^{pl}} H_{11} \approx \frac{v_{13}^{el}}{v_{31}^{el}} H_{11}.$$

$$(5.57)$$

The elastic Poisson's ratios can be directly derived from the elastic stiffness tensor determined in section 5.2. The values in expression (5.57) depend on the value of H_{11}. Application of the equations (5.48) and (5.49) leads to

$$H_{11} = \frac{h^2}{\left(Y_1^t\right)^2} = 1.$$

$$(5.58)$$

For the determination of shear relevant parameters HOFFARTH et al., 2017 proposes to fit the shear curves with the effective stress-strain master curve derived from the tensile experiment by considering the relations (5.48) and (5.49) as well as (5.50), respectively (5.52). In the present work, it is suggested to approximate the parameters by applying solely equations (5.48) and (5.49). Hence, using the initial yield values, the shear relevant parameters of the flow function are determined by

$$H_{44} = \frac{h^2}{\left(Y_{23}^s\right)^2} = \frac{\left(Y_1^t\right)^2}{\left(Y_{23}^s\right)^2},$$

$$H_{55} = \frac{h^2}{\left(Y_{13}^s\right)^2} = \frac{\left(Y_1^t\right)^2}{\left(Y_{13}^s\right)^2},$$

$$H_{66} = \frac{h^2}{\left(Y_{12}^s\right)^2} = \frac{\left(Y_1^t\right)^2}{\left(Y_{12}^s\right)^2}.$$

$$(5.59)$$

The approach presented in this section is proposed to allow for a simplified parametrisation of the material model. The full validity of this approach in respect to different matrix materials and fibre contents was not checked in this work. In case a more exact distinction of elastic and plastic Poisson's ratio is necessary, the material model could be easily extended to enable the additional input of plastic Poisson's ratio values.

5.3.8 Approximation of Hardening Curves

In this section an analytical approximation of the experimentally acquired plastic hardening curves is proposed. The experimental hardening curve for uniaxial tension is obtained by relating plastic strains, determined by equation (3.51), to the related stress values.

An analytical approximation of the hardening curve has the benefit of offering a simple description of plastic behaviour without requiring the information of every stress-strain point of the experimental curve. This allows for a simple input of hardening information to the material model.

The parametrised analytical hardening curve avoids the complicated preconditioning of input curves, as discussed e.g. in VOGLER, 2014. Furthermore, derivatives of the hardening curve in respect to the internal variable, required in (5.56) can be readily found.

In the following, an analytical expression shall be derived that makes use of the material parameters previously determined by micro-mechanical expressions. This expression is a function of plastic strain ε^{pl}. It is based on the evaluation of the plastic secant compliance S^{pl} of the material, which is defined by

$$S^{pl}\left(\varepsilon^{pl}\right) = \frac{\varepsilon^{pl}}{\sigma - \sigma_Y}. \tag{5.60}$$

Equation (5.60) represents all the secants determinable between arbitrary plastic strain-stress points and the initial plastic strain-stress point at yield σ_Y for the inverse hardening curve of a material. Since equation (5.60) is undefined at the point $\sigma = \sigma_Y$, respectively $\varepsilon^{pl} = 0$, this point is excluded from the evaluation. At this point the following relation holds:

$$\sigma\left(\varepsilon^{pl} = 0\right) = \sigma_Y. \tag{5.61}$$

The evaluation of the plastic secant compliance in order to determine the hardening behaviour was adopted from PAETOW, 1991 who applied a similar method for paper, as described in PFEIFFER et al., 2019. SCHMACHTENBERG, 1985 proposed an analogous procedure to describe the stress curve of amorphous and semi-crystalline non-reinforced thermoplastics in relation to total strain, cf. OSSWALD et al., 2006.

In figure 5.11, plastic compliance curves were derived from tensile experiments with specimens extracted from different PPGF30 source parts by application of expression (5.60). It shows that these curves can be approximated by a linear fit. By introducing modelling parameters ζ_0 and ζ_1, this leads to expression

$$S^{pl}\left(\varepsilon^{pl}\right) = \zeta_1\varepsilon^{pl} + \zeta_0. \tag{5.62}$$

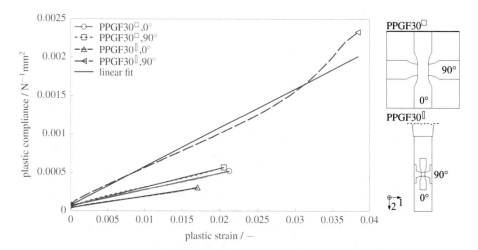

Figure 5.11 Plastic compliance curves for tensile tests of PPGF30.

Rearranging (5.60) and inserting (5.62) yields

$$\sigma(\varepsilon^{pl}) = \frac{\varepsilon^{pl}}{\zeta_1 \varepsilon^{pl} + \zeta_0} + \sigma_Y. \tag{5.63}$$

This expression essentially represents an inverse polynomial as defined by NELDER, 1966. Recently, DRASS et al., 2018 proposed the application of the Nelder polynomial for modelling polymeric adhesives. In the further, suggestions for the determination of the parameters ζ_0 and ζ_1 are given.

Figure 5.11 shows that the parameter ζ_1 represents the slope of the linear fit. The final values of the curve are given by the plastic failure strain ε_F^{pl} and the stress at failure σ_F. The initial value of the curve at $\varepsilon^{pl} = 0$ is given by ζ_0. From these considerations the hardening parameter ζ_1 is defined as

$$\zeta_1 = \frac{\frac{\varepsilon_F^{pl}}{\sigma_F - \sigma_Y} - \zeta_0}{\varepsilon_F^{pl}} = \frac{1}{\sigma_F - \sigma_Y} - \frac{\zeta_0}{\varepsilon_F^{pl}}. \tag{5.64}$$

The derivative of equation (5.60) can be expressed as

$$\frac{\partial \sigma}{\partial \varepsilon^{pl}} = \frac{-\varepsilon^{pl} E}{\left(\varepsilon^{pl} E + \zeta_0\right)^2} + \frac{1}{\varepsilon^{pl} E + \zeta_0}. \tag{5.65}$$

The initial slope of the hardening curve gives

$$\frac{\partial \sigma}{\partial \varepsilon^{\text{pl}}}(0) = \frac{1}{\zeta_0}. \tag{5.66}$$

In the following, it is discussed how the initial hardening slope and therefore the parameter ζ_0 can be expressed. In the linear elastic region, the stress-strain curve is modelled by a linear slope defined by the Young's modulus E. However, as described in section 4.3.2.4, the experimental stress-strain curve shows non-linear characteristics in this region. In order to evaluate yield stress, an offset strain value ε_{rp} was introduced in this context. A parallel line to the linear elastic line, offset by the parameter ε_{rp} from the origin strain, cuts the non-linear stress-strain curve at the yield point.

It is assumed that the non-linear stress-strain slope in the elastic region can be adequately expressed by the quadratic expression

$$\sigma^{\text{qdr}} = \zeta_1^{\text{qdr}}\varepsilon + \zeta_2^{\text{qdr}}\varepsilon^2. \tag{5.67}$$

Variables related to the quadratic expression are labelled by the superscript qdr. The parameters ζ_1^{qdr} and ζ_2^{qdr} of the quadratic expression are derived in the following.

Relation (5.67) implies that the quadratic curve passes through the origin. The derivative of the quadratic curve in respect to strain at this point is defined by the Young's modulus. This leads to

$$\frac{\partial \sigma^{\text{qdr}}}{\partial \varepsilon}(0) = \zeta_1^{\text{qdr}} = E. \tag{5.68}$$

By definition, the quadratic curve passes through the yield point σ_Y at the strain value $\varepsilon_Y = \varepsilon_{\text{rp}} + \sigma_Y/E$. Therefore, the second parameter can be expressed by

$$\zeta_2^{\text{qdr}} = -\frac{E\varepsilon_{\text{rp}}}{\left(\varepsilon_{\text{rp}} + \frac{\sigma_Y}{E}\right)^2}. \tag{5.69}$$

Expression (5.67) can be related to plastic strain by inserting the expression for the strain values given in (3.51). Moreover, it has to be ensured that the expression resolves to yield stress when a plastic strain of the value zero is inserted:

$$\sigma^{\text{qdr}} = E\left(\varepsilon^{\text{pl}} + \frac{\sigma^{\text{qdr}}}{E} + \varepsilon_{\text{rp}}\right) - \frac{E\varepsilon_{\text{rp}}}{\left(\varepsilon_{\text{rp}} + \frac{\sigma_Y}{E}\right)^2}\left(\varepsilon^{\text{pl}} + \frac{\sigma^{\text{qdr}}}{E} + \varepsilon_{\text{rp}}\right)^2. \tag{5.70}$$

Solving equation (5.71) for σ^{qdr} and ensuring $\sigma^{\text{qdr}}\left(\varepsilon^{\text{pl}} = 0\right) = \sigma_Y$ gives

$$\sigma^{\text{qdr}} = -E(\varepsilon^{\text{pl}} + \varepsilon_{\text{rp}}) + \frac{\sqrt{\varepsilon_{\text{rp}}\left(\varepsilon^{\text{pl}} + \varepsilon_{\text{rp}}\right)\left(E\varepsilon_{\text{rp}} + \sigma_Y\right)^2}}{\varepsilon_{\text{rp}}}. \tag{5.71}$$

The positive solution of the derivative of (5.71) for a plastic strain value of zero is expressed as

$$\frac{\partial \sigma^{\text{qdr}}}{\partial \varepsilon^{\text{pl}}}(0) = -\frac{E - \frac{\sigma_Y}{\varepsilon_{\text{rp}}}}{2}. \tag{5.72}$$

It is assumed that the initial slope of the hardening curve, expressed by the Nelder function in (5.66), can be set equal to expression (5.72). ζ_0 is then designated by

$$\zeta_0 = -\frac{2}{E - \frac{\sigma_Y}{\varepsilon_{\text{rp}}}}. \tag{5.73}$$

Applying expression (5.64) and (5.73), the analytical hardening curve is expressed by

$$\sigma(\varepsilon^{\text{pl}}) = \frac{\varepsilon^{\text{pl}}}{\zeta_1 \varepsilon^{\text{pl}} + \zeta_0} + \sigma_Y = \frac{\varepsilon^{\text{pl}}}{\left(\frac{1}{\sigma_F - \sigma_Y} - \frac{2}{\varepsilon_F^{\text{pl}}\left(E - \frac{\sigma_Y}{\varepsilon_{\text{rp}}}\right)}\right)\varepsilon^{\text{pl}} - \frac{2}{E - \frac{\sigma_Y}{\varepsilon_{\text{rp}}}}} + \sigma_Y \tag{5.74}$$

In addition to the values E and σ_Y, equation (5.74) requires three parameters for the description of the hardening behaviour. In order to further reduced the amount of required parameters, an alternative description of ζ_1 is suggested. Considering a maximum stress limit $\sigma_{\max}$ for equation (5.63) leads to

$$\sigma_{\max} = \lim_{\varepsilon^{\text{pl}} \to \infty} \sigma\left(\varepsilon^{\text{pl}}\right) = \sigma_Y + \frac{1}{\zeta_1}. \tag{5.75}$$

Hence, the value $\sigma_{\max}$ represents the asymptotic maximum value of the stress-strain curve.

Taking into account expression (5.75) and (5.73), the final analytical hardening curve is given as

$$\sigma\left(\varepsilon^{\text{pl}}\right) = \frac{\varepsilon^{\text{pl}}}{\zeta_1 \varepsilon^{\text{pl}} + \zeta_0} + \sigma_Y = \frac{\varepsilon^{\text{pl}}}{\frac{1}{\sigma_{\max} - \sigma_Y}\varepsilon^{\text{pl}} - \frac{2}{E - \frac{\sigma_Y}{\varepsilon_{\text{rp}}}}} + \sigma_Y \tag{5.76}$$

In this context, the parameters $\sigma_{\max}$ and ε_{rp} can be interpreted as fitting parameters that depend on the orientation and state of anisotropy for each tensile hardening curve.

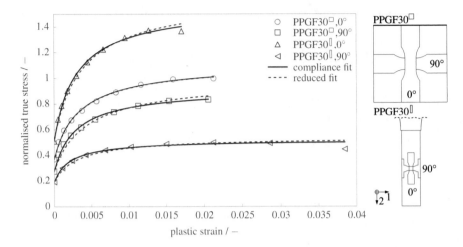

Figure 5.12 Hardening curves for tensile tests of PPGF30 derived by compliance approximation. The compliance fit is achieved by applying equation (5.76). The reduced fit assumes a constant ε_{rp} for all hardening curves.

Alongside these two modelling parameters, the yield stress and Young's modulus are needed for the complete description of the hardening curve.

The parameters have to follow the restrictions

$$\sigma_{max} > \sigma_Y \tag{5.77}$$

and

$$0 < \varepsilon_{rp} < \frac{\sigma_Y}{E}. \tag{5.78}$$

In figure 5.12, the approximation of tensile hardening curves is shown by the compliance fit lines which have been generated by applying equation (5.76). Overall, the analytical formulation gives an adequate approximation of the hardening curves. Figure 5.11 shows that the parameter ζ_0 is in a similar range for all hardening curves. The results depicted as reduced fit in figure 5.12 show that a reasonable approximation can be achieved even when ε_{rp} is set to a constant value for all hardening curves. For the presented PPGF30, the parameter of the reference curve is used, having a value of 0.001 056. The parameters E and σ_Y can be computed by applying the micro-mechanical formulations given in sections 5.2 and 5.3.1. This means that the hardening curve for every orientation state that deviates from a reference state can be reasonably derived by the variation of a single modelling parameter ζ_1, respectively σ_{max}.

The determination of the parameters was merely presented for the tensile hardening curves in this section. It has to be equally done for the remaining compressive, shear and equi-biaxial stress states.

In literature, alternative analytical descriptions for experimental hardening curves of plastics can be found. Some examples of such expressions for the depiction of stress hardening during uniaxial tensile loading are given in the following.

The original Schmachtenberg equation derived by SCHMACHTENBERG, 1985 is defined in terms of total strain as

$$\sigma(\varepsilon) = E\varepsilon \frac{1 - \zeta_1^S \varepsilon}{1 - \zeta_2^S \varepsilon}. \tag{5.79}$$

For semi-crystalline materials SCHMACHTENBERG, 1985 suggests the to set $\zeta_1^S = 0$. The resulting expression of (5.79) can be modified to represent the plastic hardening curve as

$$\sigma\left(\varepsilon^{pl}\right) = \zeta_0^S E \varepsilon^{pl} \frac{1}{1 - \zeta_2^S \varepsilon^{pl}} + \sigma_Y = \varepsilon^{pl} \frac{1}{\frac{1}{\zeta_0^S E} - \frac{\zeta_2^S}{\zeta_0^S E} \varepsilon^{pl}} + \sigma_Y. \tag{5.80}$$

This form of the Schmachtenberg expression resembles the hardening formulation derived in equation (5.76). However, the fitting parameters are defined differently.

Applying an approach by FUCHS, 2007, REITHOFER et al., 2016 suggest to adopt the (5.79) to plastic stress-strain curves by setting

$$\sigma\left(\varepsilon^{pl}\right) = \sigma_Y + E\varepsilon^{pl} \frac{1}{1 - \frac{E}{\zeta_3^S} \varepsilon^{pl}}. \tag{5.81}$$

G'SELL et al., 1979 derived an exponential expression that was used for the description of hardening of thermoplastics by e.g. BECKER et al., 2008. The G'Sell-Jonas equation for the definition of a non-linear stress-strain curve, comprising yield stress σ_Y and two modelling parameters ζ_1^{GJ} and ζ_2^{GJ}, can be modified for an approximation of the hardening curve to

$$\sigma\left(\varepsilon^{pl}\right) = \sigma_Y \left(2 - e^{-\varepsilon^{pl} \zeta_1^{GJ}}\right) e^{\zeta_2^{GJ} (\varepsilon^{pl})^2}. \tag{5.82}$$

Several authors such as e.g. SCHÖPFER, 2011 propose a simplification of the hardening slope by an exponential expression translating to a linear relation. Applying modelling parameters ζ_1^{expon} and ζ_2^{expon}, it is expressed as

$$\sigma\left(\varepsilon^{pl}\right) = \sigma_Y \left(2 - e^{-\varepsilon^{pl} \zeta_1^{expon}}\right) + \zeta_2^{expon} \varepsilon^{pl}. \tag{5.83}$$

Expression (5.83) will only allow a rough approximation of non-linear experimental hardening curves. Equation (5.81) applies a constraint on the initial slope of the plastic hardening curve that overly restricts the fitting capabilities when Young's modulus E and yield stress σ_Y are predefined. The inverse polynomial expressions from Schmachtenberg (5.80) and from the present work (5.74) and (5.76) offer similarly adequate fitting of non-linear hardening curves. Equations (5.80), (5.74) and (5.76) will offer equal results if the parameters are set accordingly. The G'Sell-Jonas expression (5.82) gives comparably reasonable fitting results for the hardening curves of polymeric materials. However, in comparison to the exponential expression of G'Sell-Jonas, the parameters of the hardening approximation obtained in (5.74) and (5.76) are defined on the basis of characteristic values of the hardening curve, which can be derived in an engineering context. Expressions (5.74) and (5.76) allow the direct input of micro-mechanically obtained Young's modulus E and yield stress σ_Y. Therefore, they can be easily applied in the context of combining the micro-mechanical and phenomenological modelling as proposed in this work. Since equation (5.76) requires less parameters than (5.74) it is applied in the further.

5.3.9 Consideration of Different Orientation Distributions

Based on the results of section 5.3.8, it is assumed that the hardening curve of an SFRTP material with an arbitrary anisotropy degree can be expressed by (5.76). In the following, a method is proposed by which the hardening relation for arbitrary orientation degrees can be approximated from the hardening curve of a reference state.

It was shown in figure 5.12 that a reasonable fit of the hardening curve can be achieved even when the parameter ζ_0 is set to a constant for all curves of one stress state. Presuming that Young's modulus and yield stress are known, this leaves the parameter σ_{max} to be determined for each orientation state.

It is assumed that the relation between σ_Y and σ_{max} is unique for every state of anisotropy. The comparative values ζ_{σ_Y} and $\zeta_{\sigma_{max}}$ are defined in relation to the reference curve as

$$\zeta_{\sigma_Y} = \frac{\sigma_Y}{\sigma_{Y,\text{reference}}} \tag{5.84}$$

and

$$\zeta_{\sigma_{max}} = \frac{\sigma_{max}}{\sigma_{max,\text{reference}}}. \tag{5.85}$$

The values defined in expressions (5.84) and (5.85) are plotted in figure 5.13 for the case of uniaxial tensile loading. σ_Y has been determined from the experiments by the method presented in section 4.3.2.4. The values for σ_{max} have been determined from fitting expression (5.76) to each experimental hardening curve by a least squares method. The

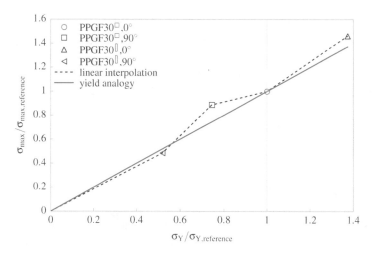

Figure 5.13 Relation of σ_Y to σ_{max}. The interpolation line connects the experimental values while the term yield analogy is based on the assumption $\frac{\sigma_{max}}{\sigma_{max,reference}} = \frac{\sigma_Y}{\sigma_{Y,reference}}$.

results from the PPGF30$^\square$ tension bar tested at a loading angle of $0°$ are used as reference. Since the $45°$ tensile test does not represent a uniaxial tensile stress state in the material coordinate system it is not included in the plot.

Applying these results, it is suggested to generate hardening curves for the orientation distributions that have not been evaluated in experiments from an interpolation of the data in figure 5.13. The value σ_Y can be determined by the micro-mechanical relations given in section 5.2 for arbitrary orientation distributions. The value σ_{max} needed for (5.76) can then be extracted from figure 5.13.

In the context of the presented approach, the value σ_Y can be interpreted as an effective value for the characterisation of a specific orientation distribution, rather than an actual physical value.

For the parametrisation of the material model, the data points needed for the interpolation are stored as tabulated values relating ζ_{σ_Y} to $\zeta_{\sigma_{max}}$.

It can be seen in figure 5.13 that a rough approximation for the values of σ_{max} is given by an overall linear assumption. For this reason, in case sufficient experimental data is not available and no tabular data can be given for the interpolation, the linear approximation is applied by the material model. This approach is termed yield analogy, since it is based on the assumption that the maximum stress values σ_{max} change analogous to the yield stress values σ_Y. This can be expressed as

$$\sigma_{max}(\sigma_Y) = \frac{\sigma_{max,reference}}{\sigma_{Y,reference}} \sigma_Y. \qquad (5.86)$$

The mapping of σ_Y to σ_{max} has to follow the constraint given in equation (5.77).

The method presented for the hardening curves from uniaxial tensile loading has to be applied equally for the additional stress states of shear, compression and equi-biaxial tension. This is necessary to determine the Tsai-Wu parameters determined in expressions (5.22) and (5.23) and to account for different hardening behaviour at each of these stress states. If the relations between yield values and hardening curve parameter σ_{max} are not given to the material model for additional stress states, the relations for uniaxial tension are used.

5.3.10 Rate Dependency

Strain rate dependency is implemented using a consistent approach, cf. section 3.7. In consistent theory, the current yield surface is assumed to be also depending on the rate of change of the internal variable in addition to the current value of the internal variable.

As discussed in section 4.4.2, it can be assumed for the tested PPGF30 material that the strain rate dependency of the normalised plastic deformation curves is approximately independent of loading direction and can be assumed to be isotropic. The experiments have shown the normalisation assumption to be appropriate for the tested PPGF30. Thus, strain rate dependency must not be considered on the matrix level but can be accounted for on the composite level. This approach is applied in the material model by scaling the composite yield surface to account for strain rate effects.

In the present work, the range of validity of this assumption was not further evaluated in experiments apart from the PPGF30 composite. Nevertheless, the micro-mechanical equations for the initial yield stress given in (5.5) respectively (5.17) theoretically indicate this effect for all possible SFRTPs. Assuming the matrix yield stress in (5.17) is increased due to strain rate effects, the composite yield surface will be scaled equally in all loading directions. In case the stress concentration tensor $\mathbb{T}$ is not affected by the strain rate, this occurs independently of the fibre orientation distribution. This condition will always be fulfilled when the Young's modulus is not influenced by the strain rate and thus viscoelastic behaviour is neglected. This is due to the stress concentration factor being independent of the yield stress as shown in section 5.2. Hence, in this case the normalisation of a composite yield value influenced by strain rate in respect to a quasi-static composite yield value will give the same results for any loading direction.

Nevertheless, strain rate dependent behaviour of actual composite materials does include viscoelastic effects. As shown in section 5.2, the stress concentration tensor is influenced by the stiffness of the matrix. In this context, the strain rate dependent behaviour of the composite will be the result of a combination of the change of Young's modulus and yield strength. As a conclusion, it could be assumed that the isotropic strain rate dependency assumption will give more adequate results if viscoelastic effects can be neglected.

As shown in section 5.3.6, hardening of the yield surface of the suggested model is implemented by a modification of the yield values it is based on. To incorporate rate dependency based on the consistent approach, equation (5.55) is extended to

$$Y = f(Y_0, \psi, \dot{\psi}). \tag{5.87}$$

This means, that additionally to the internal variable ψ, the rate of change of the internal variable $\dot{\psi}$ is considered.

Equation (5.56) can be adjusted to the strain rate related derivative by substituting $\partial \psi$ with $\partial \dot{\psi}$.

Since the internal variable is set equal to the effective plastic strain in this work, viscoplasticity is controlled by the effective plastic strain rate.

Following the experimental results in section 4.4.2, a scaling of the yield surface by a function depending on strain rate is assumed. In this context, the G'Sell-Jonas or the Johnson-Cook model as presented in section 4.3.2.6 can be applied.

While the respective equations in section 4.3.2.6 depend on the strain value, in the context of simple model parametrisation, a single parameter set for the control of strain rate dependency would be preferable. This can be realised by defining a parameter set that in average gives the best approximation of the strain rate dependent behaviour over the entire deformation range. In this case, the G'Sell-Jonas or Johnson-Cook equation are applied to scale the hardened stress value $\sigma(\varepsilon^{pl})$ from equation (5.63).

However, the results in section 4.4.2, respectively figure 4.32 demonstrate that the strain rate dependency in the region of small deformation may strongly deviate from the strain rate dependency at higher deformations. The present work proposes to take this into account by directly scaling the specific values σ_Y and σ_{max} of equation (5.63) with two separate strain rate dependent parameter sets. The scaling of the initial value $\sigma_{Y,0}$ considers the strain rate dependency at low plastic strains while the the scaling of the intial value $\sigma_{max,0}$ regards strain rate dependent behaviour at increased plastic strains. In the material model this effectively only requires one additional parameter, since the reference strain rate $\dot{\varepsilon}_0$ is set equal for both the scaling of $\sigma_{Y,0}$ and $\sigma_{max,0}$. Therefore, the overall strain rate dependent hardening curve with the G'Sell-Jonas model is defined as

$$\sigma(\varepsilon^{pl}, \dot{\varepsilon}^{pl}) = \cfrac{\varepsilon^{pl}}{\cfrac{\varepsilon^{pl}}{\sigma_{max,0}\left(\frac{\dot{\varepsilon}^{pl}}{\dot{\varepsilon}_0}\right)^{\zeta_1^{GJ}} - \sigma_{Y,0}\left(\frac{\dot{\varepsilon}^{pl}}{\dot{\varepsilon}_0}\right)^{\zeta_2^{GJ}}} - \cfrac{2}{E - \frac{\sigma_{Y,0}}{\varepsilon_{rp}}\left(\frac{\dot{\varepsilon}^{pl}}{\dot{\varepsilon}_0}\right)^{\zeta_2^{GJ}}}} + \sigma_{Y,0}\left(\frac{\dot{\varepsilon}^{pl}}{\dot{\varepsilon}_0}\right)^{\zeta_2^{GJ}} \tag{5.88}$$

with the strain rate behaviour controlling parameters ζ_1^{GJ} and ζ_2^{GJ} as well as the predefined reference strain rate $\dot{\varepsilon}_0$ for quasi-static loading.

For the Johnson-Cook model it is represented by

$$\sigma(\varepsilon^{pl}, \dot{\varepsilon}^{pl}) = \cfrac{\varepsilon^{pl}}{\cfrac{\varepsilon^{pl}}{\sigma_{max,0}-\sigma_{Y,0}+\left(\sigma_{max,0}\zeta_1^{JC}-\sigma_{Y,0}\zeta_2^{JC}\right)\ln\left(\frac{\dot{\varepsilon}^{pl}}{\dot{\varepsilon}_0}\right)} - \cfrac{2}{E-\frac{\sigma_{Y,0}}{\varepsilon_{rp}}\left(1+\zeta_2^{JC}\ln\left(\frac{\dot{\varepsilon}^{pl}}{\dot{\varepsilon}_0}\right)\right)}}$$
$$+\sigma_{Y,0}\left(1+\zeta_2^{JC}\ln\left(\frac{\dot{\varepsilon}^{pl}}{\dot{\varepsilon}_0}\right)\right)$$

$$(5.89)$$

with the strain rate behaviour controlling parameters ζ_1^{JC} and ζ_2^{JC}.

In order to to fulfil the parameter restriction given in (5.77) and ensure that the yield stress σ_Y does not surpass the maximum stress σ_{max}, the related parameters are bound by

$$\zeta_2^{GJ} < \cfrac{\ln\left(\frac{\sigma_{max,0}}{\sigma_{Y,0}}\left(\frac{\dot{\varepsilon}^{pl}}{\dot{\varepsilon}_0}\right)^{\zeta_1^{GJ}}\right)}{\ln\dot{\varepsilon}^{pl} - \ln\dot{\varepsilon}_0}$$

$$(5.90)$$

for the G'Sell-Jonas model and by

$$\zeta_2^{JC} < \cfrac{\frac{\sigma_{max,0}}{\sigma_{Y,0}}\left(1+\zeta_1^{JC}\ln\left(\frac{\dot{\varepsilon}^{pl}}{\dot{\varepsilon}_0}\right)\right)-1}{\ln\left(\frac{\dot{\varepsilon}^{pl}}{\dot{\varepsilon}_0}\right)}$$

$$(5.91)$$

for the Johnson-Cook model.

In the presented implementation the same strain rate dependency is assumed for all stress states. However, the model could be extended to consider a strain rate influence that depends on the specific stress state. DÖLL, 2016 shows that such a stress state dependency exists in thermoplastic materials. This effect could be included in the material model by defining a different set of parameters ζ_1^{JC} and ζ_2^{JC} respectively ζ_1^{GJ} and ζ_2^{GJ} for each stress state. Moreover, the authors of SCHOSSIG et al., 2008 suggest a modified G'Sell-Jonas model for short fibre reinforced PP composites with two different parameter sets for isothermal behaviour at low strain rates $<20/s$ and adiabatic behaviour at strain rates $>20/s$. This approach could be also be implemented into the proposed model but would lead to a more elaborate parametrisation process.

5.4 Modelling of Damage and Failure

5.4.1 Damage Model

It is assumed that strain values are equal in the damaged and the undamaged configuration by applying the strain equivalence expression (3.62), cf. LEMAITRE et al., 2005. This allows the plasticity and damage calculations to be uncoupled in the implemented material model. In this context, the damage implementation follows the proposed concept of HOFFARTH, 2016 who applies similar yield and potential functions as defined in the present work and shows the validation of the strain equivalence assumption.

The relation between stresses in the damaged and the undamaged system for the one dimensional case were given in expression (3.60). Considering a damage tensor $\boldsymbol{D}$ in Voigt's notation, this can be written for the 3D case as

$$\tilde{\vec{\sigma}} = \boldsymbol{D}\vec{\sigma}. \tag{5.92}$$

In the general case the damage tensor has 21 components. However, in literature many simplified proposals for the damage tensor can be found. Following a simplified approach commonly applied to composites, the damage tensor is presumed to be diagonal, e.g. MATZENMILLER et al., 1995. Direct coupling of the stress components by the damage tensor is thus not considered. For the case of isotropic damage, the expressions should ensure that the damage tensor reduces to a scalar value.

For continuous fibre reinforced composites HOFFARTH et al., 2017 propose a multiplicative approach to consider different damage values due to loading in different directions.

The damage tensor can be further simplified if it is assumed that damage is introduced in the principal stress directions only. Expressing this assumption for an arbitrary coordinate system and introducing damage variables d_1, d_2 and d_3 the following form is proposed by CHOW et al., 1987:

$$\boldsymbol{D} = \begin{bmatrix} 1-d_1 & 0 & 0 & 0 & 0 & 0 \\ 0 & 1-d_2 & 0 & 0 & 0 & 0 \\ 0 & 0 & 1-d_3 & 0 & 0 & 0 \\ 0 & 0 & 0 & \sqrt{(1-d_2)(1-d_3)} & 0 & 0 \\ 0 & 0 & 0 & 0 & \sqrt{(1-d_1)(1-d_3)} & 0 \\ 0 & 0 & 0 & 0 & 0 & \sqrt{(1-d_1)(1-d_2)} \end{bmatrix}. \tag{5.93}$$

For the short fibre composites considered in the present work, the normalisation approach presented in section 4.4.3 is applied. The experiments have shown that for tension the individual damage variables can be approximated by

$$d_1 = d_2 = d. \tag{5.94}$$

In the material model this approximation is also assumed for the remaining tensile component:

$$d_3 = d \tag{5.95}$$

By application of expressions (5.94) and (5.95), the damage tensor for the material model is assumed to be approximately isotropic and reduces to scalar damage. In the experiments this assumption has only been validated for tensile stress states. A validation of the model for alternative stress states such as shear has to be performed in further research.

The proposed model indicates an evolution of damage for all stress states. However, in the material effects such as crack closure under compression may occur, which are not considered. Therefore, the proposed model may only be adequate for simple singular loading and unloading processes. In case more complex loading processes including multiple loading cycles and more intricate loading paths are to be predicted, a more sophisticated damage model is probably required.

5.4.2 Failure

In the context of this work, the selection of failure criteria was restricted to those that are capable of describing yield or rupture of chopped short fibre reinforced composites with arbitrary fibre orientations. See section 3.5 for a short overview of failure criteria

Even though originally developed for UD composites, mechanistic expressions could be adjusted to fit failure surfaces for SFRTP materials. However, in such a case the original physical justification of these criteria would not hold. The experiments have shown that the non-ideal fibre-matrix interface influences material failure, see section 4.4.2. Since an experimental derivation of interface properties is not readily performable, the complex implementation of failure into the micro-mechanical model on micro-level is liable to a large number of unaccountabilities. A number of researchers such as e.g. PIERARD, 2006 and KAISER, 2013 have implemented failure criteria on the micro-level. Nevertheless, an iterative reverse engineering of failure parameters necessitates from such an implementation.

Considering the aforementioned drawbacks of alternative methods, the description of failure behaviour by the implementation of global phenomenological criteria on the macro-level presents a straightforward engineering approach. Global criteria intrinsically include the different failure effects on the micro-level and are more easily adjustable than complexly intertwined micro-level criteria.

The inspection of failure behaviour of SFRTP has not been the focus of this work. Therefore, regarding material failure, this work was limited to global tensor criteria. In this context, similarly to the yield criterion, a Tsai-Wu failure criterion was selected. This shall not indicate that the selected criterion offers the best description of these materials, since the target of this work was not the comparison of different failure criteria, but rather the presentation of a simplified modelling approach. The Tsai-Wu criterion regarding failure of SFRTP is discussed and successfully applied by HEGLER, 1987, SCHÖPFER, 2011, GRUBER et al., 2012, KAISER, 2013 and NUTINI et al., 2017 among others.

To allow for a simple parametrisation of the failure criterion, the assumption was made that the yield surface and the failure surface are interdependent. This interdependency is introduced by relating the failure value X_F to the yield value relation:

$$X_F = f\left(\frac{\sigma_Y}{\sigma_{Y,\text{reference}}}\right). \tag{5.96}$$

Following the method introduced in section 5.3.9 for the hardening curve, the failure values can be extracted by linear interpolation from a table that is given as input to the material model.

If sufficient experimental data points for tabulated failure values in respect to the yield relation are not available, it is assumed that the change of the failure value X_F equals the inverse of the change of σ_Y as in

$$X_F = \left(\frac{\sigma_Y}{\sigma_{Y,\text{reference}}}\right)^{-1} X_{F,\text{reference}}. \tag{5.97}$$

This method is applied separately for the derivation of failure values for uniaxial tension, shear, equi-biaxial tension and compression. This is due to the fact, that the experimental results in section 4.4.4 show an influence of the stress state on the failure strain.

Finally, all twelve failure values necessary for the derivation of the respective parameters of an orthotropic Tsai-Wu failure surface can be determined, cf. equation (3.58).

The relations previously derived can be adapted to describe failure values either in stress or in strain space. The material model allows the specification of one of these variants. In the first case, the failure value X_F is defined as failure stress σ_F. In the second case, the failure value X_F equals the plastic failure strain $\varepsilon_F^{\text{pl}}$.

The experimental results in section 4.4.2 show that failure depends on the strain rate during loading. Therefore, the failure values derived previously for the quasi-static case are scaled to meet failure values at higher strain rates. This is done similarly to the yield value scaling described in section 4.3.2.6.

The Johnson-Cook equation (4.8) can be adjusted to calculate failure values $X_F(\dot{\varepsilon})$ that depend on the strain rate $\dot{\varepsilon}$ by

$$X_F(\dot{\varepsilon}) = X_{F,0}\left(1 + \zeta_F^{\text{JC}} \ln\left(\frac{\dot{\varepsilon}}{\dot{\varepsilon}_0}\right)\right). \tag{5.98}$$

The failure parameter ζ_F^{JC} controls the strain rate dependency. $\dot{\varepsilon}_0$ represents a predefined reference strain rate at approximately quasi-static loading to which the current strain rate $\dot{\varepsilon}$ is related. $X_{F,0}$ is the failure value that occurs at the predefined strain rate $\dot{\varepsilon}_0$.

Alternatively the G'Sell-Jonas equation (4.9) can be adapted to calculate the strain rate depending failure value $X_F(\dot{\varepsilon})$ by setting

$$X_F(\dot{\varepsilon}) = X_{F,0} \left(\frac{\dot{\varepsilon}}{\dot{\varepsilon}_0} \right)^{\zeta_F^{GJ}}. \tag{5.99}$$

In this context, the strain rate dependency parameter for failure ζ_F^{GJ} needs to be defined for the material model.

In the previous equations it was assumed that the parameters ζ_F^{JC} and ζ_F^{GJ} are equal for every loading direction and stress state. However, e.g. NUTINI et al., 2017 have shown for a different PPGF30 material that the strain rate dependency of failure depends on the loading direction. This was considered in NUTINI et al., 2017 by introducing a different linear dependence on the logarithm of the strain rate for each parameter of the Tsai-Wu model. In the material model of the present work such an effect can be included by introducing a different failure parameter ζ_F^{JC} respectively ζ_F^{GJ} for each loading direction. Moreover, the parameters could also be adjusted for each stress state to further refine the prediction quality.

5.5 Numerical Implementation

The constitutive equations described in the previous sections were implemented into an explicit FEA solver. In a strain-driven simulation, as in an LS-DYNA explicit context, the task of a material model is the incremental update of element stresses, compare HALLQUIST, 2006. The constitutive model is included into the FEA simulation code as a user-subroutine. The subroutine receives information on stresses from the last calculated time-step $n-1$ and incremental strains from the current step n from the kinematic computations of the FEA software. Additional information from previous calculation steps is saved by the material routine itself in form of history variables. Furthermore, the subroutine needs access to all the necessary material parameters that are given to the simulation by an input file. With the gathered information the subroutine can calculate and update the stress values and return them to the main FEA program.

The total strain increment that is passed to the material model is assumed to be the sum of elastic and plastic strain. The incremental form of the elasto-plastic decomposition from expression (3.49) is given as

$$d\vec{\varepsilon} = d\vec{\varepsilon}^{el} + d\vec{\varepsilon}^{pl}. \tag{5.100}$$

Elastic stress is defined by Hooke's law (3.40). The range of plastic deformation is defined by a yield surface. Plastic flow arises when the yield condition in expression (3.47) is fulfilled. Yielding depends on the stress state $\boldsymbol{\sigma}$ and the internal variable ψ that governs hardening, respectively yield surface evolution. Hence, the condition of plastic flow can be expressed as

$$\Phi(\sigma, \psi) = 0. \tag{5.101}$$

In section 5.3.6, the effective plastic strain $\overline{\varepsilon}^{\mathrm{pl}}$ was defined as the single internal variable. Plastic strains are related to plastic stresses by the flow rule, see (3.52). The consistency condition ensures that no load points lay outside of the enveloping yield surface, cf. equation (3.48). It ensures that all new plastic states fulfil the yield condition. Simplification of the consistency condition to a single internal variable leads to

$$\Phi(\sigma + d\sigma, \psi + d\psi) = \Phi(\sigma, \psi) + d\Phi = 0. \tag{5.102}$$

Since Hooke's law holds in incremental form, elastic stresses can be computed after an elastic stiffness matrix has been determined by the material model. For the derivation of plastic stresses, an elastic predictor-plastic corrector approach is used. Herein, all strains are initially assumed to be elastic and the related stresses are computed. Since it is not known if the material response is actually purely elastic, this elastic predictor stress is also termed trial stress. In case the trial stress lies outside the enveloping yield surface, plasticity needs to be considered and the pre-calculated stresses need to be projected back to the yield surface. They must finally fulfil the yield as well as the consistency condition.

Using Hooke's law from (3.40), the elasto-plastic split from equation (5.100) and the flow rule from (3.52), leads to the incremental expression

$$\mathrm{d}\vec{\sigma} = \boldsymbol{C}\mathrm{d}\vec{\varepsilon}^{\mathrm{el}} = \boldsymbol{C}\left(\mathrm{d}\vec{\varepsilon} - \mathrm{d}\vec{\varepsilon}^{\mathrm{pl}}\right) = \underbrace{\boldsymbol{C}\mathrm{d}\vec{\varepsilon}}_{\text{elastic predictor}} - \underbrace{\boldsymbol{C}\mathrm{d}\overline{\lambda}\,\frac{\partial h}{\partial\vec{\sigma}}\left\|\frac{\partial h}{\partial\vec{\sigma}}\right\|^{-1}}_{\text{plastic corrector}}. \tag{5.103}$$

This expression comprises the previously described elastic predictor-plastic corrector form.

Following the strain equivalence approach, it is possible to perform all necessary material modelling calculations in the undamaged system. In the material modelling context, the damaged stresses $\tilde{\vec{\sigma}}_{n-1}$ of the previous time-step $n-1$ are transformed to the undamaged system $\vec{\sigma}_{n-1}$ by the relation given in expression 5.92. This is done prior to applying the elastic predictor-plastic corrector calculations. After all computations have been performed, the calculated stresses are re-transformed into the damaged system.

The trial stress is calculated by assuming a purely elastic behaviour:

$$\vec{\sigma}_n = \vec{\sigma}_{n-1} + \boldsymbol{C}\mathrm{d}\vec{\varepsilon}^{\mathrm{el}}. \tag{5.104}$$

For the plastic correction step, a return mapping algorithm is used which will be explained in detail in the next section.

5.5.1 Return Mapping Algorithm

The process of mapping back the stresses to the current yield surface is realised by a return mapping algorithm. A multitude of return mapping algorithms exist, cf. e.g. SIMO et al., 1998. Out of these, the cutting plane return mapping algorithm (CPRMA) as described by ORTIZ et al., 1986 was adapted to the material model in the present work. The advantage of this algorithm is that it only requires first order derivatives of the yield and flow rules. As can be seen in equation (5.56) the first order derivatives are already extensive.

The CPRMA is based on the linearisation of the consistency condition from equation (5.102) as

$$\Phi(\boldsymbol{\sigma},\boldsymbol{\psi}) + d\Phi = \Phi(\boldsymbol{\sigma},\boldsymbol{\psi}) + \frac{\partial \Phi}{\partial \vec{\sigma}} d\vec{\sigma} + \frac{\partial \Phi}{\partial \psi} d\psi = 0. \tag{5.105}$$

The CPRMA is an explicit algorithm. Multiple explicit CPRMA steps may be needed to map the initial trial stress back to the yield surface before the stress update for a time-step computation can be performed. In the following, the index m will refer to the explicit step of the CPRMA, while the index n will refer to the overall calculation step of the material subroutine. Expression (5.105) at a current material routine calculation step is adapted to the current CPRMA calculation step by

$$\begin{aligned}
0 &= \Phi_{n,m} + \frac{\partial \Phi_{n,m}}{\partial \vec{\sigma}} d\vec{\sigma}_{n,m} + \frac{\partial \Phi_{n,m}}{\partial \psi} d\psi_{n,m} \\
&= \Phi_{n,m} + \frac{\partial \Phi_{n,m}}{\partial \vec{\sigma}} (\vec{\sigma}_{n,m+1} - \vec{\sigma}_{n,m}) + \frac{\partial \Phi_{n,m}}{\partial \psi} (\psi_{n,m+1} - \psi_{n,m}).
\end{aligned} \tag{5.106}$$

In the context of this work, hardening is assumed to be a function of effective plastic strain. Hence, the internal variable equals the effective plastic strain. Adapting expression (5.45) and taking into account equation (5.49), the internal variable can be related to the plastic multiplier by

$$\begin{aligned}
d\psi_{n,m} &= d\bar{\varepsilon}_{n,m}^{\mathrm{pl}} = d\bar{\lambda}_{n,m} \frac{h_{n,m}}{\sqrt{(H\vec{\sigma})^{\mathrm{T}}(H\vec{\sigma})}} = d\bar{\lambda}_{n,m} \left\| \frac{\partial h_{n,m}}{\partial \vec{\sigma}} \right\|^{-1} \\
&= \zeta_{\mathrm{d}\lambda} d\bar{\lambda}_{n,m}.
\end{aligned} \tag{5.107}$$

The incremental effective plastic strain is mapped to the plastic multiplier by the Euclidean norm of the flow potential derivative. For a clearer visualisation, the substitution $\zeta_{\mathrm{d}\lambda} = \|\partial h/\partial \vec{\sigma}\|^{-1}$ will be made in the following expressions.

Adapting equation (5.103) to the explicit scheme, an expression for $d\vec{\sigma}_{n,m}$ is obtained as

$$
\begin{aligned}
\vec{\sigma}_{n,m+1} &= C\mathrm{d}\vec{\varepsilon}_{n,m} - C\mathrm{d}\overline{\lambda}_{n,m}\frac{\partial h_{n,m}}{\partial\vec{\sigma}}\zeta_{\mathrm{d}\lambda} = \vec{\sigma}_{n,m} - C\mathrm{d}\overline{\lambda}_{n,m}\frac{\partial h_{n,m}}{\partial\vec{\sigma}}\zeta_{\mathrm{d}\lambda} \\
&\rightarrow \mathrm{d}\vec{\sigma}_{n,m} = -C\mathrm{d}\overline{\lambda}_{n,m}\frac{\partial h_{n,m}}{\partial\vec{\sigma}}\zeta_{\mathrm{d}\lambda}.
\end{aligned}
\tag{5.108}
$$

Substituting expressions (5.108) and (5.107) into (5.106) leads to

$$
\mathrm{d}\overline{\lambda}_{n,m} = \frac{\Phi_{n,m}}{\frac{\partial\Phi_{n,m}}{\partial\vec{\sigma}}C\frac{\partial h_{n,m}}{\partial\vec{\sigma}}\zeta_{\mathrm{d}\lambda} - \frac{\partial\Phi_{n,m}}{\partial\psi}\zeta_{\mathrm{d}\lambda}}.
\tag{5.109}
$$

This set of equations can now be used in order to iteratively remap stress to the yield surface, respectively find the set of plastic strain and plastic stress that solves the yield condition.

Strain rate dependency was implemented using a consistent approach where the consistency condition is still fulfilled at all stages. Thus, the equations previously derived still hold with minor adjustments.

Viscoplastic effects are assumed to be controlled by an effective plastic strain rate. In this context, a calculation step n of the material model is interpreted as a time-step with a current time Δt. The rate of change of plastic strain in this time-step is derived by the mid-point rule

$$
\dot{\overline{\varepsilon}}^{\mathrm{pl}}_{n,m} = \zeta_{\dot{\varepsilon}}\frac{\overline{\varepsilon}^{\mathrm{pl}}_{n,m}}{\Delta t} + (1-\zeta_{\dot{\varepsilon}})\dot{\overline{\varepsilon}}^{\mathrm{pl}}_{n-1}.
\tag{5.110}
$$

It has been noted by KRIVACHY, 2007 and HAUFE et al., 2005 amongst others, that the stability of rate dependent algorithms can be increased by introducing a filtering parameter that is applied as a weighting factor $\zeta_{\dot{\varepsilon}}$ in the mid-point rule. While $\zeta_{\dot{\varepsilon}}$ can take values from 0 to 1, typically small values such as 0.05 are suggested.

The yield surface in the consistent approach additionally depends on the plastic strain rate. In case no plastic strain has been calculated yet, equation (5.110) would give a result of zero for the strain rate. This would lead to initial yielding always occurring at the quasi-static yield point, regardless of the current strain rate. Therefore, total effective strain instead of plastic effective strain is inserted in (5.110) at the initial trial step to consider a strain rate for the initial yield check.

Since in the consistent approach the yield surface is depending on plastic strain rate, the linearisation of the consistency condition from equation (5.102) now resolves to

$$
0 = \Phi_{n,m} + \frac{\partial\Phi_{n,m}}{\partial\vec{\sigma}}\mathrm{d}\vec{\sigma}_{n,m} + \frac{\partial\Phi_{n,m}}{\partial\psi}\mathrm{d}\psi_{n,m} + \frac{\partial\Phi_{n,m}}{\partial\dot{\psi}}\mathrm{d}\dot{\psi}_{n,m}.
\tag{5.111}
$$

Taking into account equation (5.111) and the previous statements for the rate-independent case, the plastic multiplier for the CPRMA is derived as

$$d\overline{\lambda}_{n,m} = \frac{\Phi_{n,m}}{\frac{\partial \Phi_{n,m}}{\partial \overline{\sigma}} C \frac{\partial h_{n,m}}{\partial \overline{\sigma}} \zeta_{d\lambda} - \frac{\partial \Phi_{n,m}}{\partial \psi} \zeta_{d\lambda} - \frac{\partial \Phi_{n,m}}{\partial \psi} \zeta_{d\lambda} \frac{1}{\Delta t}} \tag{5.112}$$

In some cases it was perceived that the algorithm would not converge when the yield surface was surpassed for the first time and strain rate influence would initially come into account. This is due to the yield surface depending on two likewise interrelated variables, thus there may exist more than one solution to the yield condition. The CPRMA may lead to oscillating results in case the consistency condition gives values under zero at an iteration step, VOGLER, 2005.

For this case, a secant iteration scheme was implemented. The secant iteration scheme enforces that the yield condition is met and oscillations are limited. The basis of the secant iteration scheme is again a linearisation of the consistency condition, but while expression (5.106) utilises derivatives in order to approximate subsequent iteration steps, the secant iteration scheme applies a secant between two subsequent yield condition values from the CPRMA as

$$0 = \Phi_{n,m-1} + \frac{\Phi_{n,m-1} - \Phi_{n,m-2}}{d\overline{\lambda}_{n,m-1} - d\overline{\lambda}_{n,m-2}} \left(d\overline{\lambda}_{n,m} - d\overline{\lambda}_{n,m-1} \right). \tag{5.113}$$

Equation (5.113) can be rearranged to determine updated values of the plastic multiplier by

$$d\overline{\lambda}_{n,m} = \Phi_{n,m-1} \frac{d\overline{\lambda}_{n,m-1} - d\overline{\lambda}_{n,m-2}}{\Phi_{n,m-1} - \Phi_{n,m-2}} + d\overline{\lambda}_{n,m-1}. \tag{5.114}$$

All equations required for the material subroutine have been presented. Given the expressions for $d\overline{\lambda}_{n,m}$ in (5.109), (5.112) or (5.114), the plastic multiplier can be approximated in multiple iterations m. In each of these iteration steps, after $d\overline{\lambda}_{n,m}$ is found, the corrector stress $d\vec{\sigma}_{n,m}$ must be updated and the newly predicted overall stress $\vec{\sigma}_{n,m+1}$ has to be computed by application of equation (5.108). Next, the effective plastic strain $\overline{\varepsilon}^{pl}_{n,m+1}$ and the effective plastic strain rate $\overline{\dot{\varepsilon}}^{pl}_{n,m+1}$ have to be determined with $d\overline{\lambda}_{n,m}$ by use of expressions (5.45) and (5.110). Finally, the yield condition $\Phi_{n,m+1}(\vec{\sigma}_{n,m+1}, \overline{\varepsilon}^{pl}_{n,m+1}, \overline{\dot{\varepsilon}}^{pl}_{n,m+1})$ can be computed. Convergence of the yield condition is checked against an adequately small tolerance value. If it is fulfilled, the previously calculated values $m+1$ are set to the current overall material model step values n. Otherwise, the values $m+1$ are set to the current iteration step values m and the iterative return mapping process restarts. After the return mapping algorithm has converged, stresses and strains need to be updated so that they can be returned to the overlaying FEA code.

The overall numerical algorithm that was implemented in the material model is briefly explained in the next section.

5.5.2 Overview of the Implementation

All of the equations given in chapter 5 were implemented in the FORTRAN 77 programming language. The user subroutine is applicable in an explicit FEA context. Algorithm 5.2 presents a detailed overview of the overall algorithm of the material model. It was implemented in the explicit framework of LS-DYNA in the context of this work.

Algorithm 5.2 Numerical implementation of the material model.

Input:
- time-step, stress, strain: $n, \vec{\sigma}_{n-1}, \mathrm{d}\vec{\varepsilon}_n$
- parameters: material constants, modelling properties
- history variables (user input): orientation tensor $\boldsymbol{A}$
- history variables (calculated): initially calculated material properties, $\mathrm{d}\dot{\bar{\varepsilon}}^{\mathrm{pl}}_{n-1}, \mathrm{d}\bar{\varepsilon}^{\mathrm{pl}}_{n-1}$

if $n \equiv 0$ **then**   /* *initial time-step: pre-calculation of elastoplastic material parameters* */

     determination of Eshelby's tensor

     micro-mechanical modelling of UD pseudo grain using Mori-Tanaka approach

     compute composite stiffness $\boldsymbol{C}$ from orientation averaging of pseudo grains

     compute yield values and yield function parameters

     compute flow potential parameters

     compute failure surface parameters

     save all derived parameters as history variables

     $\vdots$

Algorithm 5.2 (Continued)   Numerical implementation of the material model.

> $\vdots$
>
> **else if** $n > 0$ **then**   /* not initial time-step */
>> read data calculated at initial step from history variables
>> transform from damaged into undamaged system: $\vec{\sigma}_{n-1} = \tilde{\vec{\sigma}}_{n-1}/(1-d_n)$
>> set elastic trial stress $\vec{\sigma}_{n,0} = \vec{\sigma}_{n-1} + \boldsymbol{C}\mathrm{d}\vec{\varepsilon}^{\text{el}}$ and initial effective plastic strain $\overline{\varepsilon}^{\text{pl}}_{n,0} = \overline{\varepsilon}^{\text{pl}}_{n-1}$
>> relate effective plastic strain to plastic strains   /* for the evolution of hardening */
>> set $m = 0$, $m_{max} = M$, $s = 0$, $s_{max} = S$, tol$= 10^{-5}$, exit=False
>>
>> **repeat**   /* return mapping iteration loop */
>>> ▶ ITERATE
>>> set effective plastic strain rate
>>> /* check yield condition   */
>>> **if** $\Phi(\vec{\sigma}_{n,m}, \overline{\varepsilon}^{\text{pl}}_{n,m}, \dot{\overline{\varepsilon}}^{\text{pl}}_{n,m})) > $ tol **then**   /* cutting plane algorithm */
>>>> set iteration step $m = m+1$
>>>> **if** $m \geq M$ **then** go to STOP   /* plasticity algorithm failed to converge */
>>>> set yield surface derivatives $\frac{\partial \Phi_{n,m}}{\partial \vec{\sigma}}$, $\frac{\partial \Phi_{n,m}}{\partial \overline{\varepsilon}^{\text{pl}}}$, $\frac{\partial \Phi_{n,m}}{\partial \dot{\overline{\varepsilon}}^{\text{pl}}}$
>>>> set flow rule and derivatives $h(\vec{\sigma}_{n,m})$, $\frac{\partial h_{n,m}}{\partial \vec{\sigma}}$
>>>> set plastic multiplier $\mathrm{d}\overline{\lambda}_{n,m} = \dfrac{\Phi_{n,m}}{\frac{\partial \Phi_{n,m}}{\partial \vec{\sigma}} \boldsymbol{C} \frac{\partial h_{n,m}}{\partial \vec{\sigma}} \zeta_{\mathrm{d}\lambda} - \frac{\partial \Phi_{n,m}}{\partial \overline{\psi}} \zeta_{\mathrm{d}\lambda} - \frac{\partial \Phi_{n,m}}{\partial \dot{\overline{\psi}}} \zeta_{\mathrm{d}\lambda} \frac{1}{\Delta t}}$
>>>> go to UPDATE
>>> **end**
>>> **else if** abs$(\Phi(\vec{\sigma}_{n,m}, \overline{\varepsilon}^{\text{pl}}_{n,m}, \dot{\overline{\varepsilon}}^{\text{pl}}_{n,m})) \geq$ tol **or** $m > 0$ **then**   /* secant method */
>>>> set secant iteration step $s = s+1$
>>>> **if** $s \geq S$ **then** goto STOP   /* plasticity algorithm failed to converge */
>>>> **if** sgn$(\Phi_{n,m-1}) \equiv$ sgn$(\Phi_{n,m-1})$ **then** $\Phi_{n,m-1} = \Phi_{n,m-2}$ and $\Phi_{n,m-2} = \Phi_{n,m-3}$
>>>> set $\mathrm{d}\overline{\lambda}_{n,m} = \Phi_{n,m-1}\dfrac{\mathrm{d}\lambda_{n,m-1} - \mathrm{d}\lambda_{n,m-2}}{\Phi_{n,m-1} - \Phi_{n,m-2}} + \mathrm{d}\lambda_{n,m-1}$
>>>> go to UPDATE
>>> **end**
>>> **else**
>>>> go to STOP   /* if $m \equiv 0$ then elastic state, else converged plastic state */
>>> **end**
>>> ▶ UPDATE
>>> $\vec{\sigma}_{n,m+1} = \vec{\sigma}_{n,m} - \boldsymbol{C}\mathrm{d}\lambda_{n,m}\frac{\partial h_{n,m}}{\partial \vec{\sigma}}\zeta_{\mathrm{d}\lambda}$ and $\overline{\varepsilon}^{\text{pl}}_{n,m+1} = \overline{\varepsilon}^{\text{pl}}_{n,m} + \mathrm{d}\lambda_{n,m}\zeta_{\mathrm{d}\lambda}$
>>> set $\vec{\sigma}_{n,m} = \vec{\sigma}_{n,m+1}$ and $\overline{\varepsilon}^{\text{pl}}_{n,m} = \overline{\varepsilon}^{\text{pl}}_{n,m+1}$
>>> go to ITERATE   /* restart loop */
>>> ▶ STOP
>>> set final stress $\vec{\sigma}_n = \vec{\sigma}_{n,m}$ and strain $\overline{\varepsilon}^{\text{pl}}_n = \overline{\varepsilon}^{\text{pl}}_{n,m}$
>>> set exit = True   /* leave iteration loop */
>> **until** exit $\equiv$ True
>>
>> check failure condition
>> transform from undamaged into damaged system $\tilde{\vec{\sigma}}_n = \vec{\sigma}_n(1-d_n)$
>> **return** updated values $\vec{\sigma}_n$, $\overline{\varepsilon}^{\text{pl}}_n$ and updated parameters as history variables
> **end**

With algorithm 5.2, the effective linear elastic as well as the plastic SFRTP composite response can be calculated in an FEA simulation.

In case of elasticity, the input of stiffness properties of the fibres and the matrix as well as volume fractions and an average fibre aspect ratio are necessary. The average aspect ratio is used, since the fibre length distribution is not accounted for in process simulations, cf. section 2.5. In order to make the material accessible to results from process simulations, the input of a second order orientation tensor is required as an information on fibre orientation. The fourth order orientation tensor is calculated internally by the use of closure approximations, cf. section 5.2. The five individual components of the orientation tensor have to be given for each integration point of the model. They are imported from the history variables by the algorithm. Therefore, the history variables of the integration points have to be predefined in an input file.[1]

Numerical efficiency is increased by computing the linear elastic composite stiffness matrix only upon initialisation and saving the results as history variables for further simulation steps.

For plastic behaviour, the yield points of the matrix for uniaxial tension and compression, shear and biaxial tension have to be input. Furthermore, the hardening curve of the reference composite material has to be parametrised. Data on strain rate dependency, damage and failure also have to be derived from a reference SFRTP reaction and needs to be input into the material model.

---

[1] In the FEA software LS-DYNA the history variable definition depends on the element type of the simulation, e.g. an INITIAL_STRESS_SHELL or INITIAL_STRESS_SOLID card has to be used, cf. LS-DYNA, 2017.

6 Verification

In the following, the verification of the constitutive equations determined in chapter 5 is performed. The parameters used for the material modelling of the characterised PPGF30 material are given. In the first section, the stiffness predictions of the model are checked against experimental results. The second section focusses on the computation of the initial yield strength values of the material. In the final section, the complete numerically determined response of the material model is compared to the respective experiments for different stress states and orientation distributions.

6.1 Verification of Linear Elastic Response

For the experimental validation of the linear elastic model, the micro-mechanical homogenisation procedure is summarised as follows:

(1) Stiffness properties of the thermoplastic matrix and the reinforcement fibres were determined. In the case of PPGF30, the thermoplastic properties were evaluated from the experimental results of the pure matrix material. Fibre properties were extracted from literature.

(2) Fibre length was taken into account by dividing the sample into 20 slices of equal height and assuming an average fibre aspect ratio for each slice, cf. figure 4.14, *(b)*.

(3) Second and fourth order orientation tensors were calculated for each slice directly from the results of the fibre analysis. Thus, no closure approximation was necessary for the derivation of the fourth order orientation tensor.

(4) The slice volumes were homogenised separately by the Mori-Tanaka approach (5.12) and the orientation averaging procedure (5.13). This involved the virtual division of the slice volume into pseudo grains of equal orientation. Furthermore, it comprised the individual Mori-Tanaka homogenisation of the pseudo grains and subsequent orientation averaging of the whole slice volume by Voigt's approach.

(5) The effective stiffness of the entire sample was evaluated by applying Voigt's averaging over all slices.

This procedure follows a similar approach applied by the author in DILLENBERGER, 2014, and MÜLLER et al., 2015. However, the experimental data has been re-evaluated by

© Springer Fachmedien Wiesbaden GmbH, part of Springer Nature 2020
F. Dillenberger, *On the anisotropic plastic behaviour of short fibre reinforced thermoplastics and its description by phenomenological material modelling*, Mechanik, Werkstoffe und Konstruktion im Bauwesen 53, https://doi.org/10.1007/978-3-658-28199-1_6

the methods presented in chapter 4. For this reason, the data given in the following may slightly deviate from the values given in the respective works.

The arithmetic averages of the experimental values for the Young's modulus and the Poisson's ratio of the PP matrix material used in the PPGF30 composite are shown in table 6.1.

Table 6.1 Average experimental values for the Young's modulus and the Poisson's ratio of PP. Additionally, the respective simple standard deviations are given.

	Young's modulus $\mathrm{N/mm^2}$	Poisson's ratio –
PP, 0°	1721 ± 91	0.364 ± 0.025
PP, 90°	1747 ± 55	0.345 ± 0.014

The values were evaluated by applying the methods presented in section 4.3.2.4. For every testing configuration, the arithmetic averages were computed from the specific results extracted from each single test, rather than from the average stress-strain curve. The simple standard deviation of the Young's modulus is $<6\,\%$. The standard deviation of the Poisson's ratio was evaluated as $<7\,\%$. As has already been stated in section 4.4.1, the difference between samples loaded in 0° and 90° is negligible.

The elastic modelling parameters are summarised in table 6.2. The elastic matrix parameters for the micro-mechanical model were defined from the mean value of the 0° and 90° results. For the glass fibres, the values that are given in STOMMEL et al., 2018 were used.

Table 6.2 Elasticity parameters for the constituents of PPGF30.

	Young's modulus $\mathrm{N/mm^2}$	Poisson's ratio –	
PP	1734	0.35	from experiments
Glass fibre	72 000	0.18	STOMMEL et al., 2018

The material model requires the input of the fibre volume fraction. As discussed in chapter 4, the composite has a fibre weight fraction $\vartheta^f_{\mathrm{mass}}$ of 30 %. With the density of the matrix ρ^m and of the fibres ρ^f, the fibre volume fraction ϑ^f is determined by

$$\vartheta^f = \frac{1}{1 + \frac{1 - \vartheta^f_{\mathrm{mass}}}{\vartheta^f_{\mathrm{mass}}} \frac{\rho^f}{\rho^m}} . \qquad (6.1)$$

Utilising the values $\rho^m = 0.91\,\text{g/cm}^3$ provided in DOMININGHAUS, 2008 and $\rho^f = 2.55\,\text{g/cm}^3$ given in STOMMEL et al., 2018, the fibre volume fraction results to $\vartheta^f = 13.265\,\%$.

The fibre distribution properties applied in the homogenisation process are given in table 6.3 for each of the considered composite parts. The average fibre orientation tensor and the average fibre aspect ratio of the entire µCT sample volume are listed. These present the average values of the 20 slices of equal height that have been used for the homogenisation.

Table 6.3 Model parameters for glass fibre distributions in PPGF30 parts. The data is given as an average value over the entire volume. The fourth order orientation tensor is given in Voigt's notation.

Orientation tensor A						Aspect ratio φ
PPGF30$^\square$	$\begin{bmatrix} 0.4925 & 0.1013 & 0.0127 & 0.0012 & 0.0077 & 0.0281 \\ 0.1013 & 0.2473 & 0.0105 & 0.0038 & 0.0013 & 0.0262 \\ 0.0127 & 0.0105 & 0.0112 & 0.0004 & 0.0004 & 0.0007 \\ 0.0012 & 0.0038 & 0.0004 & 0.0105 & 0.0007 & 0.0013 \\ 0.0077 & 0.0013 & 0.0004 & 0.0007 & 0.0127 & 0.0012 \\ 0.0281 & 0.0262 & 0.0007 & 0.0013 & 0.0012 & 0.1013 \end{bmatrix}$					18.41
PPGF30$^\parallel$	$\begin{bmatrix} 0.7980 & 0.0557 & 0.0130 & -0.0039 & 0.0090 & 0.0001 \\ 0.0557 & 0.0461 & 0.0046 & -0.0022 & 0.0005 & 0.0003 \\ 0.0130 & 0.0046 & 0.0093 & -0.0005 & 0.0006 & -0.0002 \\ -0.0039 & -0.0022 & -0.0005 & 0.0046 & -0.0002 & 0.0005 \\ 0.0090 & 0.0005 & 0.0006 & -0.0002 & 0.0130 & -0.0039 \\ 0.0001 & 0.0003 & -0.0002 & 0.0005 & -0.0039 & 0.0557 \end{bmatrix}$					21.18

After the homogenisation procedure regarding linear elastic composite properties has been completed, the composite stiffness result in form of a tensor in Voigt's notation can be rotated into arbitrary orientations by use of special transformation matrices that are derived in TING, 1996. From each of these rotated stiffness tensors specific engineering constants can be evaluated by application of expression (3.42). If this is done for a sufficient amount of spatial orientations, a smooth 3D surface is generated which visualises the engineering constants value for all orientations. Figure 6.1 and 6.2 show such spherical projections of the Young's modulus for PPGF30$^\square$ and PPGF30$^\parallel$ parts. The surfaces are normalised to their respective maximum value. It can be observed that the increased anisotropy in the flat bar source part leads to a more pronounced ellipsoidal form of the projection compared to that of the plate source part.

These visualisations can only present the qualitative difference of the two source parts. In order to compare the results with experimental data, the aforementioned projection is restricted to planar orientations in the x-y plane of the source parts. Hence, the 3D body reduces to a curve, as seen in figure 6.3.

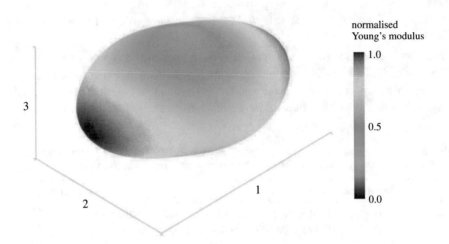

Figure 6.1 Spherical projection of Young's modulus of PPGF30$^\square$. The data is normalised in relation to the maximum Young's modulus value.

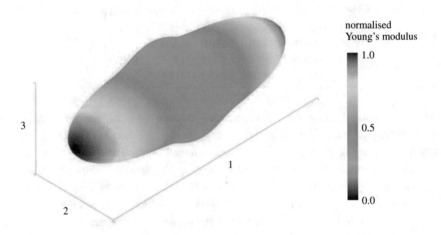

Figure 6.2 Spherical projection of Young's modulus of PPGF30$^\square$. The data is normalised in relation to the maximum Young's modulus value.

Besides visualising the results of the constitutive models with parameters derived from actual material data, 6.3 shows the results for a virtual UD and a virtual isotropic case. All

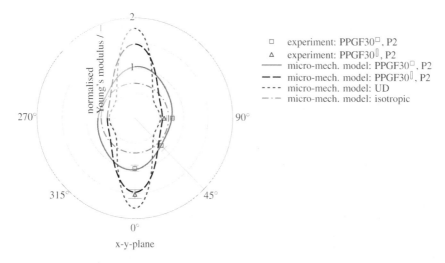

Figure 6.3   Polar plot of micro-mechanical results for the Young's modulus of PPGF30. The data is normalised in relation to the experimental value of PPGF30$^\square$ in 0°. Average experimental values are plotted with the double standard deviations.

data is normalised to the respective value of the reference test extracted from the PPGF30$^\square$ part in 0° direction.

In the case of isotropically distributed fibres, the model behaviour is equal to a circle in the polar plot, whereas it becomes a dented ellipsoid in the UD case. The Young's modulus is highest in longitudinal or 0° direction. The UD case clearly shows that material stiffness at angles between 0° and 90° can be less than at 90°. This is due to the influence of the shear modulus in these orientations. This clarifies that modelling techniques that are based on a simple interpolative approximation between behaviour in longitudinal and lateral orientation may lead to misleading assumptions.

Next to the curves derived from the material model presented in chapter 5, experimental results in specific orientations are plotted as discrete points. The model gives reasonable approximations of the stiffness values derived in experiments with the different source parts made of PPGF30.

The quantitative experimental values of Young's modulus and shear modulus of the PPGF30 parts are summarised in table 6.4. The procedures shown in section 4.3.2.4 were applied to evaluate these values. The arithmetic averages were computed from the results of each single test for every testing configuration, instead of utilising the average stress-strain curves. The simple standard deviations of the Young's moduli experimentally derived from the PPGF30 parts are <6 %. The biggest scatter of results was found for the 45° PPGF30$^\square$ sample, while the 90° PPGF30$^\square$ sample showed the smallest deviation of <2 %.

Table 6.4 Average experimental values for the modulus of PPGF30 parts under tensile and shear loading. Additionally, the respective simple standard deviations are given.

Experiment	Young's modulus N/mm²	Shear modulus N/mm²
PPGF30□, 0°	4433 ± 138	1359 ± 81
PPGF30□, 45°	3414 ± 199	1516 ± 39
PPGF30□, 90°	3425 ± 66	-
PPGF30▯, 0°	6720 ± 188	1034 ± 61
PPGF30▯, 90°	2675 ± 112	1058 ± 58

Similarly to the values for the Young's modulus, the results for the shear modulus of the different PPGF30 parts are plotted in figure 6.4. The results for PPGF30□ closely match the experimental values. In case of PPGF30▯, the model deviates from the average experimental values by approximately 12 %. However, the values are within the range of the double standard deviation of the experiments. Figure 6.4 shows that shear modulus can be adequately represented by the constitutive equations.

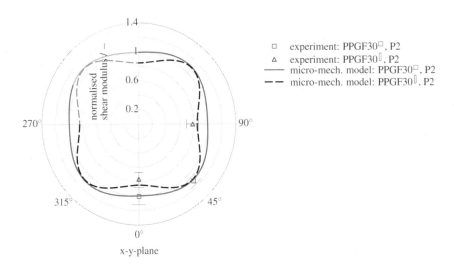

Figure 6.4 Polar plot of micro-mechanical results for shear modulus of PPGF30. The data is normalised in relation to the experimental value of PPGF30□ in 0°. Average experimental values are plotted with the double standard deviations.

The explicit analytical equations for the Mori-Tanaka model have been adopted from QUI et al., 1990 for the inclusion of transversely isotropic reinforcement fibres in the present work, cf. section 5.2 and appendix B.2. In order to check the performance of the material model regarding these types of fibres, experimental data of carbon fibre reinforced PP evaluated in AMBERG et al., 2013 was compared to the modelled stiffness data. The results are shown in appendix C, in figure C.1. A reasonable approximation quality of the model was found for the considered materials.

Overall, it can be stated that the micro-mechanical equations implemented for linear-elasticity can adequately approximate the experimental values.

6.1.1 Influences on Model Predictions

When a material model is used for the prediction of component behaviour, it is necessary to know how the variance of experimental data used for parametrisation influences the model response. Such a consideration will give an approximative overview of the models robustness in an application context. As shown in chapter 5, the entire model is built up on the basis of micro-mechanical homogenisation of the linear elastic stiffness. Therefore, the influence of test data quality on the stiffness response of the model shall be evaluated in the following.

Figure 6.5 shows the percentage deviation of the longitudinal Young's modulus of the PPGF30$^\square$ part in relation to a reference value. In order to compute the reference value, the parameter data given in the previous section was applied, cf. tables 6.3 and 6.2. The stiffness data deviating from the reference value was computed with the proposed micro-mechanical equations by varying these input parameters. The respective amount of adjustment to a specific input parameter i is specified on the axis labelled as "variation of value i" in figure 6.5. To determine each curve in the figure, the specified parameter i was varied, while all other parameters were kept constant. Among the varied values are material parameters such as the stiffness ratio of matrix to fibres and the fibre volume percentage. Moreover, fibre analysis parameters such as the fibre aspect ratio and the amount of small fibres evaluated in the micro-mechanical homogenisation process are regarded.

The largest influence can be clearly ascribed to the material parameter values. The variance of the ratio of stiffness or the volume percentage lead to a similar change of the overall results. While volume percentage has the largest influence on the stiffness predictions, it can generally be most precisely predicted. The volume percentage can be derived form the mass percentage of fibres in the composite. The mass percentage can be precisely kept during the injection moulding production process. This has been shown for example by the results of the thermogravimetric analysis in chapter 4.3.3.1. However, during form filling segregation of fibres and thermoplastic melt may occur locally due to flow conditions. These inhomogeneities concerning local volume fraction of fibres cannot be predicted by injection moulding simulation tools to date.

Stiffness of the thermoplastic bulk and hence the stiffness ratio of fibres and matrix is affected by multiple aspects. It is influenced by environmental and ageing conditions, as has been discussed in chapter 2.1. Furthermore, injection moulding process settings,

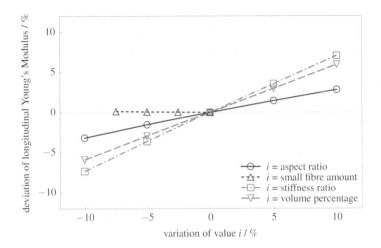

Figure 6.5 Influences on the longitudinal Young's modulus predictions of the material model.

such as mould temperature, mass temperature of the melt and pressure conditions among others, have a distinct effect on mechanical material properties, as has been discussed e.g. by KUNKEL, 2018. Thermoplastic properties may even vary between different material batches.

Since fibre-matrix interface properties are not evaluated explicitly by the material model, they are regarded implicitly by adjusted mechanical characteristics of the thermoplastic matrix permeated by fibres in the composite setting.

Considering fibre analysis parameters, the value depicted as small fibre amount is related to the break-off point of the fibre analysis algorithm. The algorithm will interpret fibres of a certain minimal length as image noise and omit these from the final evaluation, GLÖCKNER et al., 2016. Presumptions regarding the influence of the amount of small fibres taken into account for overall stiffness calculations were made in MÜLLER, 2015, but the graph clearly shows that this influence is negligible in a certain range. This is at least the case when the overall mean aspect ratio is not changed by fibre omission during analysis.

The variance of aspect ratio has a distinct influence on the computed stiffness values. A variation of the aspect ratio by $\pm 10\,\%$ will lead to a deviation of stiffness values of $\pm 4\,\%$.

It can be concluded that a complex multitude of influences on the thermoplastic matrix properties exists. Therefore, generally a determination of the composite matrix properties through an iterative reverse engineering process is more practicable than the predetermination of these properties. Reverse engineering has to be performed based on the evaluation of tests conducted on a reference experiment with the reinforced material.

6.2 Verification of Plastic Response

As shown in section 5.3.1, the micro-mechanical results for linear elasticity determined in section 6.1 have an influence on the yield surface computation. Additionally, matrix yield data is required. This data was determined in an iterative process, involving comparison of the model's responses with results from experiments in specific orientations. The experimental results from PPGF30$^\square$ were used as reference. The experimental values from PP were utilised as initial optimisation parameters.

The procedure introduced in 4.3.2.4 was applied to determine experimental yield stress data under tension and shear.

Regarding the values for compression this process did not lead to consistent results, since the initial stiffness of the test results is prone to high scatter due to e.g. friction effects, progression of multiaxial stress states or the occurrence of instabilities. These effects have been discussed in chapter 4.4.4. Therefore, a different approach was used. In this context, the relation of yield stresses from compression and tension was assumed to be approximately equal to the relation of the first stress maxima. This leads to the relation

$$\sigma_Y^c = \frac{\sigma_{max}^c}{\sigma_{max}^t} \sigma_Y^t. \tag{6.2}$$

The standard deviations for the calculated yield values from compression were determined by applying the Gaussian error propagation law, see GEVATTER et al., 2006.

The results of the experimentally and iteratively derived yield values for the matrix material are summarised in table 6.5.

Table 6.5 Average experimental values for yield strength of PP. Additionally, the respective simple standard deviations are given.

	Yield strength		
	Tension N/mm^2	Compression N/mm^2	Shear N/mm^2
PP, experiment	10.24 ± 1.20	11.84 ± 1.40	5.48 ± 0.51
PP, model	10.14	11.77	5.87

The arithmetic averages were computed from the values of every single test for each testing configuration. The yield values evaluated from the experiments show a higher standard deviation than the Young's modulus values. For the PP material, the standard deviation is in the range of <12 %. It can be seen that the yield values determined by the experiments with the pure PP material correspond to the values for the material model that were computed by the iterative process.

Application of this matrix material data to the constitutive equations leads to results for the yield values of different PPGF30 parts. To ensure that the proposed method does not only lead to an adequate yield surface fit of the reference experiment but gives reasonable approximations for all extraction directions and different source parts, the model's results need to be compared to the experimental data given in table 6.6.

Table 6.6 Average experimental values for yield strength of PPGF30 under tensile, compressive and shear loading. Additionally, the respective simple standard deviations are given.

| Experiment | Yield strength | | |
	Tension N/mm^2	Compression N/mm^2	Shear N/mm^2
PPGF30$^\square$, 0°	24.08 ± 2.51	28.07 ± 3.07	9.33 ± 1.93
PPGF30$^\square$, 45°	19.81 ± 1.59	-	10.35 ± 1.21
PPGF30$^\square$, 90°	17.95 ± 1.91	23.90 ± 2.62	-
PPGF30$^|$, 0°	33.09 ± 1.75	34.79 ± 2.17	9.09 ± 1.29
PPGF30$^|$, 90°	12.49 ± 0.80	19.81 ± 1.32	7.65 ± 1.25

The standard deviations shown in table 6.6 for the tensile and compressive yield values of the PPGF30$^\square$ are in the range of 10 %. However, the PPGF30$^|$ parts shows more reproducible results with the standard deviation of the yield values being <7 %. The scatter of the shear yield values is in a higher range of <20 % for the composite parts considered.

In figure 6.6 the material model is compared to the experimental yield values from tensile loading. The figure shows that the model predictions are in the area of the experimental values.

The values for compressive loading are compared to the experiments in figure 6.7. The modelled results of PPGF30$^\square$ closely correspond to the experimental values in both tested orientations. However, the results of PPGF30$^|$, while qualitatively resembling the experimental relations, deviate up to almost 15 % from the tested averages. With these differences, the results of the model just lay outside of the range of the double standard deviations from the experiments.

Overall, the results discussed in this section verify a reasonable prediction by the constitutive equations regarding the initial composite yield surface.

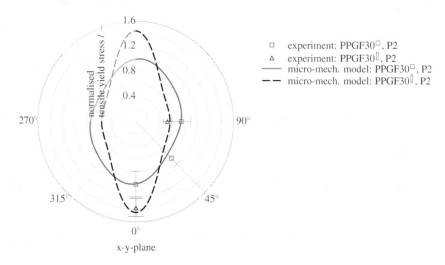

Figure 6.6   Polar plot of yield data for tension derived from micro-mechanical model of PPGF30. The data is normalised in relation to the experimental value of PPGF30$^\square$ in 0°. Average experimental values are plotted with the double standard deviations.

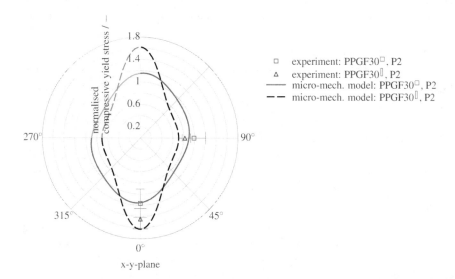

Figure 6.7   Polar plot of yield data for compression derived from micro-mechanical model of PPGF30. The data is normalised in relation to the experimental value from tension of PPGF30$^\square$ in 0°. Average experimental values are plotted with the double standard deviations.

6.3 Numerical Verification of the Material Model

6.3.1 Overview

In order to verify the non-linear plastic behaviour of the material model, the kinematic relations for the different loading cases under consideration were evaluated and the constitutive equations given in chapter 5 were solved.

Basically, the derived constitutive equations can be realised in any explicit FEA software. However, a full FEA was not the topic of the present work and the performance of the formulation was not verified in a full FEA framework. In the context of this work, the material model was implemented in the explicit code of LS-DYNA merely to derive the necessary kinematic computations. An evaluation of effects such as e.g. influences of element definitions, mesh dependency or numerical performance in terms of computational speed was hence not performed. This could be the topic of future research. Further details regarding FEA can be found in e.g. STEINKE, 2007, ZIENKIEWICZ et al., 2005 and STOMMEL et al., 2018.

As presented in the previous chapter, the material model is divided into two parts. All of the micro-mechanical calculations for the derivation of the stiffness tensor and the initial yield surface are performed in the initial time-step. The results of these complex computations are saved and retrieved in the subsequent time-steps. Hence, in all post-initial time-steps only the computations based on the phenomenological modelling need to be performed. A summary of the overall numerical procedure is given in figures 6.8 and 6.9 for the example of uniaxial tensile loading of PPGF30$^\square$ in 0°.

The evaluation of failure criteria was not the focus of this work. Consequently, only a single fracture criterion was considered. As presented in chapter 5.4, a Tsai-Wu based limit surface in stress space was applied which has been determined by a scaling of the initial yield surface. However, the fitting of a stress based failure criterion is problematic for materials that show non-linear plastic behaviour such as the PPGF30 under consideration. Such a failure definition is not stable, since the onset of failure strongly depends on the form of the hardening curve. In the special case of a hardening curve that leads to a non-increasing stress curve, the fracture criterion may not be reached. Here, a strain based failure criterion would be more stable since strain always increases with growing deformation. In such a case, the defined failure point would always be reached. Thus, a strain based failure approach is proposed for future research. It has to be noted that the applied failure criterion was also not considered for the full size specimen. Therefore, the applied failure criterion only gives a rough approximation of the actual failure on specimen level.

As basis for the phenomenological model, the micro-mechanical results of the PPGF30 composite for arbitrary orientations need to be derived in the initial time-step. These have been evaluated in the previous sections. The results of the phenomenological non-linear model computations are compared to the experimentally determined stress-strain curves from chapter 4 in the present section.

The parameters relevant for defining the hardening curve and the failure surface are given in table 6.7. The parameters for σ_{max} and σ_F are given in form of interpolated values in respect to σ_Y, as described in section 5.3.9. Here, the values of the interpolation curves are normalised in respect to the reference yield stress $\sigma_{Y,reference}$. The reference stress is determined from tensile loading of PPGF30$^\square$ in 0°. The values for σ_{max} and σ_F under equi-biaxial tension were not specifically evaluated and the respective values were copied from uniaxial tension. The failure values for uniaxial compression were set so that failure will effectively not occur in this stress state, compare section 6.3.4.

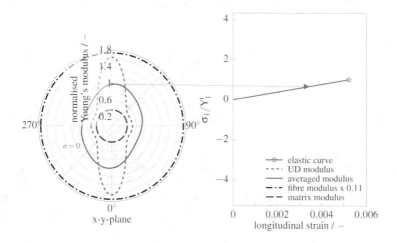

Figure 6.8 Overview of elastic modelling, exemplary for tension of PPGF30$^\square$ samples in 0°. The figure visualises the stiffness tensor by displaying the Young's modulus for arbitrary orientations in the x-y-plane. The matrix modulus and the fibre modulus as well as the fibre volume fraction are applied to derive the UD modulus by micro-mechanical modelling. The fibre modulus in the graph has been scaled down for visualisation purposes. The constitutive model utilises orientation averaging in order to include the orientation distribution. The computation ot the averaged composite stiffness tensor is exclusively done in the initial time-step $n = 0$. In subsequent time-steps, the averaged modulus is then used to compute the linear elastic deformation.

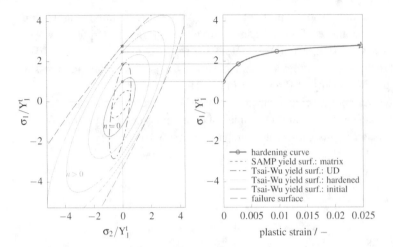

Figure 6.9 Overview of plastic modelling, exemplary for tension of PPGF30$^{\square}$ samples in 0°. The limit surfaces are displayed in stress space in the $\sigma_1 - \sigma_2$ plane. In the initial time-step a Tsai-Wu UD composite yield surface is derived from the SAMP matrix yield surface by micro-mechanical modelling. The actual orientation distribution is included by orientation averaging. Thereby, the Tsai-Wu yield surface of the composite is computed for the initial time step $n = 0$. In the subsequent time-steps $n > 0$, non-linear plastic deformation is considered by isotropic hardening based on the scaling of the initially computed Tsai-Wu yield surface. A Tsai-Wu based failure surface defines the limit of possible deformation.

Table 6.7 Parameters for hardening and failure of PPGF30. The values for σ_{max} and σ_F are interpolated in relation to σ_Y and normalised in respect to $\sigma_{Y,reference}$. Biaxial values have not been evaluated and are taken from uniaxial tension. Failure in uniaxial compression is set to not occur.

	Values							
Parameters	Tension		Shear		Compression		Biaxial Tension	
$\sigma_{Y,reference}$	$23.94\,\mathrm{N/mm^2}$							
ε_{rp}	0.000759		0.00246		0.000759		0.000759	
	$\dfrac{\sigma_Y}{\sigma_{Y,reference}}$	$\dfrac{\sigma_{max}}{\sigma_{Y,reference}}$	$\dfrac{\sigma_Y}{\sigma_{Y,reference}}$	$\dfrac{\sigma_{max}}{\sigma_{Y,reference}}$	$\dfrac{\sigma_Y}{\sigma_{Y,reference}}$	$\dfrac{\sigma_{max}}{\sigma_{Y,reference}}$	$\dfrac{\sigma_Y}{\sigma_{Y,reference}}$	$\dfrac{\sigma_{max}}{\sigma_{Y,reference}}$
σ_{max}	0.000	0.000	0.000	0.000	0.000	0.000	0.000	0.000
(interpolation	0.564	1.398	0.672	2.100	0.714	2.150	0.564	1.398
in relation	0.747	2.536	0.788	2.600	0.903	3.000	0.747	2.536
to σ_Y)	1.000	3.029			1.159	3.200	1.000	3.029
	1.455	4.292			1.629	4.000	1.455	4.292
	$\dfrac{\sigma_Y}{\sigma_{Y,reference}}$	$\dfrac{\sigma_F}{\sigma_{Y,reference}}$	$\dfrac{\sigma_Y}{\sigma_{Y,reference}}$	$\dfrac{\sigma_F}{\sigma_{Y,reference}}$	$\dfrac{\sigma_Y}{\sigma_{Y,reference}}$	$\dfrac{\sigma_F}{\sigma_{Y,reference}}$	$\dfrac{\sigma_Y}{\sigma_{Y,reference}}$	$\dfrac{\sigma_F}{\sigma_{Y,reference}}$
σ_F	0.000	0.000	0.000	0.000	0.000	0.000	0.000	0.000
(interpolation	0.564	1.373	0.672	1.253	0.714	2.172	0.564	1.388
in relation	0.747	2.306	0.788	1.399	0.903	3.091	0.747	2.318
to σ_Y)	1.000	2.778			1.159	3.258	1.000	2.778
	1.455	3.864			1.629	4.010	1.455	3.864

6.3.2 Tensile Loading

Figure 6.10 shows that the experimental results for tensile loading of the PPGF30$^\square$ plate can be reproduced for 0°, 45° and 90° loading. In the case of the 45° test, this is only possible due to the capability of the material model to adjust the behaviour for stress states which differ from tension. The increased failure strain in the 45° tension test results from the influence of shear components of the failure surface.

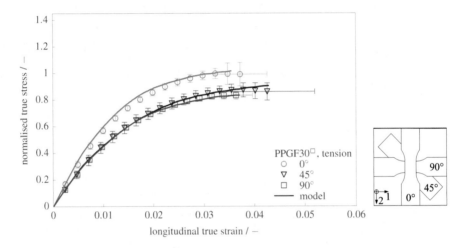

Figure 6.10   Verification results compared to tensile experiments of PPGF30$^\square$.

The interpolation procedure introduced in section 5.3.9 allows for precise fitting of the model behaviour to arbitrary orientation distributions. Figure 6.11 shows that this yields an adequate approximation for the behaviour of the PPGF30$^\lozenge$ samples with a higher orientated fibre distribution than the PPGF30$^\square$ samples.

Table 6.8 presents the parameters controlling strain rate dependent behaviour. These where determined to best fit the experimental data.

Table 6.8   Strain rate dependency parameters for hardening and failure of PPGF30.

Parameters	Values
$\dot{\varepsilon}_0$	0.0004
ζ_1^{GJ}	0.050
ζ_2^{GJ}	0.034
ζ_F^{GJ}	0.034

The strain rate dependent stress-strain curves are given for 0° tensile loading of PPGF30$^\square$ samples in figure 6.12. The depicted curves at constant strain rates were de-

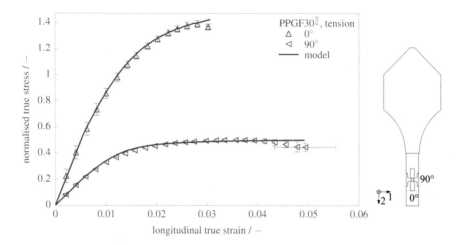

Figure 6.11 Verification results compared to tensile experiments of PPGF30▯.

termined form the experimental stress-strain curves by the procedure discussed in section 4.3.2.6. The values 0.0004/s, 3/s, 30/s and 100/s approximately correspond to the mean strain rates of the experiments at the testing velocities 1.0 mm/min, 0.1 m/s, 1 m/s and 3 m/s.

For low strain rates, the material model adequately represents the actual material behaviour. However, for high strain rates it underestimates the experimentally determined behaviour. This result occurs due to the fact that viscoelastic behaviour was not implemented. Upon regarding the initial slopes of the stress-strain curves in 6.12, a strain rate influence is already visible in the region of elastic deformation. This influence leads to a deviation between the experiments and the results of the constitutive equations at small deformations that then continues with increasing deformation.

SCHÖPFER, 2011 proposes a modification of the material modulus for material models that do not include viscoelastic behaviour in order to adjust the model's response for relevant strain rates. Such an approach was followed in figure 6.13. The Young's modulus was adjusted so as to provide for a more adequate modelling of stress-strain behaviour at high strain rates. Since all non-linear parts will be automatically adjusted due to the formulation of the hardening curve, this procedure is directly realisable in the proposed model. This is a benefit in comparison to material models that rely on hardening curves that are input as tabulated incremental data. In such a case, all hardening curves need to be readjusted when the material modulus is modified, since they include the initial yield values.

In order to include damage effects into the material model, an assumed undamaged hardening curve was applied. During the deformation process, the stresses were then

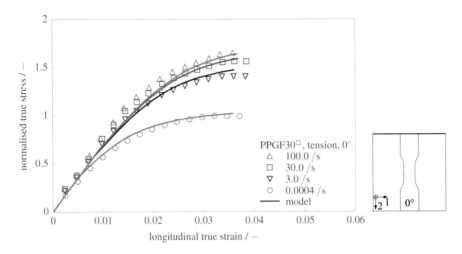

Figure 6.12 Strain rate dependent behaviour of the material model compared to tensile experiments of PPGF30□. Elastic behaviour was adjusted to the quasi-static curve.

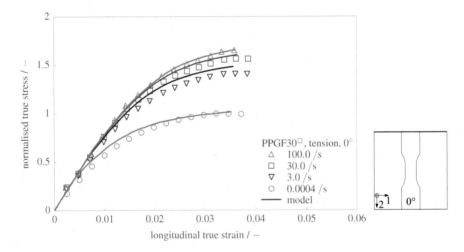

Figure 6.13 Strain rate dependent behaviour of the material model compared to tensile experiments of PPGF30□. Elastic behaviour was adjusted to the curve at highest strain rate.

adjusted by the damage relations given in section 5.4. The relevant damage parameters are given in table 6.9.

The resulting stress-strain curve for cyclic tensile loading in 0° is compared to the experimental results in figure 6.14. It can be seen that the damage modelling results in a modulus reduction which adequately represents the effect that was seen during the experiments. As presented in section 4.4.3, viscose effects are neglected. Hence, the overall

Table 6.9 Adjusted modelling parameters for the consideration of damage of PPGF30.

Parameters	Values for Tension	
ε_{rp}	0.000 244	
$\tilde{\zeta}_0$	0.129 89	
$\tilde{\zeta}_1$	3874.849	
	$\dfrac{\sigma_Y}{\sigma_{Y,\text{reference}}}$	$\dfrac{\sigma_{\text{max}}}{\sigma_{Y,\text{reference}}}$
	0.000	0.000
σ_{max}	0.564	2.879
(interpolation)	0.747	5.222
	1.000	6.237
	1.455	8.837

non-linear elastic unloading respectively reloading curves are approximated by a linear slope.

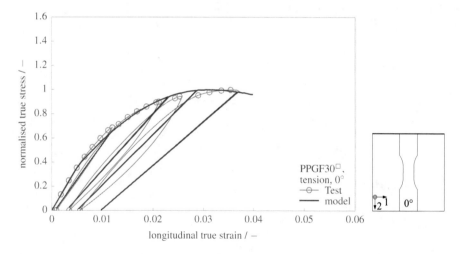

Figure 6.14 Results for cyclic tensile loading compared to PPGF30$^\square$ experiment.

6.3.3 Shear Loading

Figure 6.15 presents the results of the model for shear loading. The qualitative characteristics of the experimental stress-strain curves are reproduced by the material model. However, while the elastic deformation is adequately represented in the linear region, the response deviates from the experimental values with increasing deformation. The maximum stress values are overestimated. In this context, it is necessary to re-evaluate the actual specimens from the experiments.

Figure 6.16, *a* shows that shear strain localises in the shear zone during deformation. At the beginning of the deformation process, a simple shear state exists in the experiments and the model reproduces the experimental curves. Upon increasing deformation, the shear specimen slightly rotates in the clamping unit. With ongoing deformation, tension is superimposed on the sample and a tensile strain concentrates in the radius region, see figure 6.16, *b*. This eventually leads to tensile fracture of the specimens as shown in figure 6.16, *c*. Moreover, it has been discussed by e.g. HAUFE et al., 2010 and LIN et al., 2013 for different simple shear test samples that in regions of larger deformation the stress triaxiality corresponds to a mixed stress state of shear and tension. Accordingly, the presented shear verification test cannot adequately represent the experimental stress state.

Overall, it is clear from these results that the applied shear sample is not suitable for the short fibre reinforced materials under consideration. It is not possible to derive values for shear fracture with these samples. Alternative samples have been evaluated for example by KAISER, 2013, where similar effects of superposition of shear loading and tensile loading were observed. As a conclusion, the shear failure parameters introduced in the present work were derived so that the failure surface adequately fits the tensile test in 45° and were not determined from the shear tests.

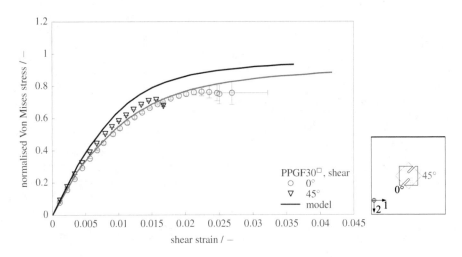

Figure 6.15 Verification results compared to shear experiments of PPGF30$^{\square}$.

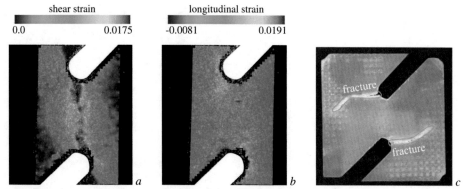

Figure 6.16   Deformation of $0°$ PPGF30$^\square$ shear samples: *(a)* DIC results for shear strain at average shear strain of 0.013, *(b)* DIC results for longitudinal strain at average shear strain of 0.013, *(c)* Tension induced fracture of the shear sample.

6.3.4 Compressive Loading

As seen in figure 6.17, the material model was calibrated so as to approximately reproduce the maximum stresses under compressive loading. As described in section 4.4.4, inhomogeneous deformation and instabilities occurred during compression testing. The material failure could not be evaluated for the compression samples. For this reason, failure was not considered in the material model.

It is shown that the model results do not include the softening that is observed in the experiments of most of the test configurations after the maximum stress is reached. The detailed inspection of the sample deformation under compression performed in section 4.4.4 gives an explanation for this softening effect. It can be ascribed to the strongly inhomogeneous deformation that occurs in the samples after the stress maximum. The deviation of model curve and experimental curve results from the fact that the verification test does not consider the inhomogeneous deformation thereafter. An uniaxial compressive stress state is only present up to the point of the stress maximum in the experiments. However, in the case of the PPGF30$^\square$ sample loaded in $90°$, the model adequately represents the experimental stress-strain curve. It can be seen in figure 4.48 that the samples were homogeneously deformed and, therefore, a uniaxial stress state is assumed in this case.

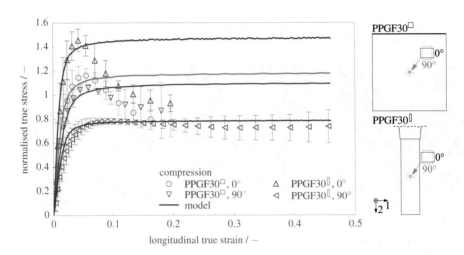

Figure 6.17   Verification results compared to compression experiments of PPGF30$^\square$ and PPGF30$^\parallel$.

7 Considerations for Engineering Practice

In engineering practice, the product development process is mostly divided into consecutive design steps that require varying degrees of detail regarding the prediction of the mechanical behaviour of parts.

In the last steps of the product development process, the engineering goal is to optimise and finalise the part design. This process requires a most precise prediction of material behaviour in order to allow for an optimal utilisation of the materials load bearing capabilities. In the majority of technical applications which consider SFRTPs, the full utilisation of their lightweight potential is particularly significant. In safety relevant applications, an adequate prediction of deformation capabilities, respectively, energy absorption and failure is crucial. In this context, a detailed experimental characterisation and a constitutive model that allows for a precise description of the material behaviour are required.

In contrast, in early pre-design steps, the target is to give an approximative prediction of part behaviour in order to e.g. enable the pre-calculation of part and material costs or the estimation of assembly space. In such a stage of the design process experimental data is mostly unavailable, respectively, budget for the derivation of material data is strongly limited. Furthermore, a large experimental evaluation of material properties and complex material models that require extensive parametrisation are impossible to apply due to strict time constraints. Nevertheless, the engineer requires mechanical material properties in order to predict material behaviour.

In the previous chapters, the capabilities of the constitutive model have been discussed regarding a detailed description of SFRTPs. The model allows for an adequate prediction of the mechanical behaviour which is necessary in the final steps of the product design process. However, the presented model also encompasses simplifications that allow for a less complex parametrisation of the constitutive equations. These simplifications can be applied in early design steps. The present chapter summarises these aspects and complements the suggestions for the determination of material parameters for anisotropic viscoplastic SFRTPs in such a context. It gives recommendations on the required experiments and furthermore suggests the application of the constitutive equations presented in this work for the parametrisation of existing material models in engineering practice.

© Springer Fachmedien Wiesbaden GmbH, part of Springer Nature 2020
F. Dillenberger, *On the anisotropic plastic behaviour of short fibre reinforced thermoplastics and its description by phenomenological material modelling*, Mechanik, Werkstoffe und Konstruktion im Bauwesen 53, https://doi.org/10.1007/978-3-658-28199-1_7

7.1 Determination of the Elastic Stiffness Matrix

The linear elastic SFRTP behaviour can be determined by combining the analytical micro-mechanical equations (B.14) and the orientation averaging expression (5.14). These equations can be computed using standard mathematical software packages. The results can then be applied to determine the composites stiffness response in an arbitrary orientation or to parametrise the orthotropic stiffness matrix of an available material model in an FEA software.

The determination of the composite stiffness requires information on the fibre stiffness, the stiffness of the thermoplastic polymer, the FOD and the fibre aspect ratios. The fibre stiffness can be extracted from literature, c.f. for example SCHÜRMANN, 2007. The exact thermoplastic matrix properties are usually not available. Thus, a reverse engineering process is required to determine them. The minimum experiments required for such a reverse engineering process are:

- A uniaxial tensile test with a specimen extracted in the direction of the principal fibre orientation axis must be carried out.

- An evaluation of the FOD in the specimen should be performed, e.g. by μCT analysis.

- Either a determination of the fibre aspect ratios or an additional tensile test transverse to the principal axis of the fibre orientation tensor should be conducted.

In case the fibre orientation and the fibre aspect ratio are both available, the elasticity parameters of the matrix need to be incrementally fitted until the computed composite stiffness response corresponds to the measured modulus of the first tensile test. When the fibre length is not accessible, it also needs to be computed in the reverse engineering process. In this context, the fibre aspect ratio, in addition to the fibre orientation, controls the difference of the composite stiffness response in the principal fibre orientation directions. Hence, during reverse engineering, firstly, the fibre aspect ratio has to be adjusted so that the relation of the computed composite stiffness in the directions considered in the two tensile experiments corresponds to the test results. Secondly, the matrix fitting has to be performed as described before.

In the majority of cases, merely results from standard tension bars are available to engineers. These tension bars mostly show a high degree of fibre orientation as is shown in AMBERG, 2004. However, the fibre orientation in SFRTP parts mostly is not unidirectional. For this reason, it is recommended to perform the aforementioned reverse engineering process with experiments that have been acquired from samples with an FOD that more closely resembles that of the part, rather than relying on the standard tension bar. The plate part presented in this work can be applied in this context.

Details on the prediction of FODs during engineering design are given in e.g. MÖNNICH, 2015, STOMMEL et al., 2018 and FORNOFF et al., 2018. Since this topic was not the focus of the present work, the following paragraphs only give an overview over this subject.

Generally, the proper description of the material anisotropy in SFRTP parts is only possible with the knowledge of the local FOD in these components. The fully integrative simulation approach allows for the consideration of FODs by coupling the structural simulation with a preceding injection moulding simulation which predicts the fibre orientation. However, this approach requires a complex calibration and validation of the individual simulation steps and the iterative coupling of these. It lacks flexibility and needs an already existing mould and process design in order to give precise predictions of the FOD. Moreover, prototypes must be available to fully validate the simulation results. Thus, the integrative simulation approach is practically only accessible in the final phase of the engineering design process.

For this reason, in early design stages with no access to information on the FOD of the component, the material behaviour is mostly described isotropically and the FOD is ignored, compare STOMMEL et al., 2018. Isotropic material parameters for this approach can be derived from the micro-mechanically determined composite properties by assuming that no fibres are oriented in through-thickness direction and the in-plane fibres are distributed isotropically. Despite this simplification, it may be possible to obtain acceptable prediction results for compact components, cf. section 2.6.

As soon as basic pre-simulations or assumptions on the principal flow directions during injection moulding are available, the anisotropic material behaviour can be considered. One possibility is to assume a homogeneous FOD throughout the component and thus only take into account the variance of the principal fibre orientation while ignoring deviations of the anisotropy degree. This approach may lead to acceptable results in mainly homogeneously flat parts. Recently, FORNOFF et al., 2018 presented a phenomenological approach for the assessment of inhomogeneous fibre distributions based on the evaluation of typical FODs in specific geometrical elements. It was evaluated that typical geometrical elements such as ribs, domes and flat regions show specific FODs which depend on the material properties, the principal flow direction and the part thickness. The FODs of these specific geometrical elements can be experimentally determined in sample parts and transferred to component segments with similar appearance. Thus, a fully precise injection moulding simulation of the component can be avoided.

7.2 Determination of the Limit of Linear Deformation

A detailed guide on the derivation of elastic and plastic parameters from experimentally acquired stress-strain curves is given in section 4.3.2.4. The stress limit of linear deformation can be determined by the application of the offset strain value introduced in section 4.3.2.4. As described this stress value is referred to as yield stress of the SFRTP, although it does not represent yielding in a strictly physical sense. The deformation is considered to be elastic up to this point. When large non-linear deformations are to be avoided this limit stress can be applied as critical design value.

The yield stress for an arbitrary stress state can be computed by the application of expression (5.54). This expression requires the parameters of the Tsai-Wu yield surface to be determined for the composite. Expression (5.19) allows for the analytical computation of the yield values of a unidirectional composite for arbitrary stress states. These values can be used to parametrise a Tsai-Wu yield surface according to expressions (5.22) and (5.23). The FOD needs to be taken into account for the composite yield surface by applying the orientation averaging process as given in expressions (5.13) and (5.26). All of these computations can be performed with standard mathematical software.

The determination of the SFRTP yield surface by application of expression (5.19) essentially requires the definition of a matrix yield surface as given for example in equation (5.15). Similarly to the acquisition of the matrix stiffness values in section 7.1, the matrix yield parameters need to be determined from a reverse engineering process. In addition to the experiments presented in 7.1, the following tests may be required:

- In case shear properties are relevant, an off-axis tensile test with a specimen extracted in 45° to the principal axis of the fibre orientation is suggested. Alternatively, if an appropriate shear specimen is available, a 0° shear test can be performed.

- In case compression properties are relevant, a 90° compression test must be carried out.

- An equi-biaxial test must be conducted, if the respective properties are required.

When only the tensile properties of the composite are relevant, the tests presented in section 7.1 are sufficient.

For the PP matrix material of the PPGF30 characterised in this work, the tensile yield strength is related to the compressive yield strength by the factor 1.16 and to the shear yield strength by the factor $1/\sqrt{3}$. Thus, considering the shear yield stress, a von Mises assumption may be sufficient. These relations have to be checked for different composites.

During reverse engineering, the parameters of the matrix yield surface have to be adjusted until the micro-mechanically computed composite yield stress values correspond to the experimental results.

The computed yield strengths can be applied as limit values for the design of components. Considering the PPGF30 tested in this work, the tensile yield strength approximately represents an average of 35 % of the bearable stress before failure. The actual fraction depends on the orientation distribution and the loading direction. If the part design is based on strain limits, the strain at yield ranges between 30 % and 50 % of the strain at failure. 30 % hold true for samples with unidirectional fibre orientation loaded in a 0° direction and 50 % are reached for the same samples loaded in 90°. For the samples extracted from plate parts, the strain at yield approximately represents an average of 36 % of the failure strain.

Parametrisation of a Hill type yield surface

As described in section 2.6, mostly a Hill yield surface is applied in common FEA simulation tools such as LS-DYNA to account for orthotropic SFRTP materials. This is due to the fact that almost all FEA software already includes material models with Hill type yield formulations. However, the implemented material models based on Hill's yield formulation lack the capabilities of the proposed novel model such as e.g. the consideration of pressure dependency and the consideration of arbitrary orientation distributions. Nevertheless, in an engineering context, the equations proposed in this work can also be applied for the parametrisation of an already implemented Hill material model for specific orientation distributions. This would be necessary when the proposed Tsai-Wu yield surface based material model is not implemented in the considered FEA software.

As a special case, the implemented material model includes a yield surface similar to Hill's yield surface definition (3.57). This applies when the pressure dependent terms in the definition of the Tsai-Wu yield condition (3.58) are omitted. The modelling equations derived in section 5.3.1 allow for the analytical determination of yield surface parameters from the matrix and fibre properties as well as given orientation distributions. The Hill type yield surface can be derived when pressure dependency is neglected in the matrix yield surface definition in section 5.3.1. This occurs when for example a von Mises formulation is used instead of the proposed SAMP formulation.

In order to apply the Tsai-Wu parameters determined by the proposed modelling equations to a material model based on a Hill yield formulation, they have to be converted into the required Hill parameters. The related equations are given in appendix B.3.

7.3 Determination of Plastic Deformation

The inverse polynomial equation (5.76) applies for the description of the non-linear plastic behaviour of SFRTP composites.

The maximum von Mises stress of the $0°$ shear sample from the PPGF30$^\square$ part tested in the present work lies at a fraction of 0.78 of the tensile stress maximum. For compression this ratio is at about 1.17. Hence, the plastic hardening curve has to be defined for each stress state.

As discussed in section 5.3.9, the non-linear plastic behaviour differs for each fibre orientation state. Hence, an interpolation of the parameters of expression (5.76) over the considered range of orientation distributions is presented. Since the interpolation is performed in relation to the yield values determined in section 7.2, these yield values can also be interpreted as effective parameters that specifically identify a particular orientation distribution. The following tests in addition to the experiments proposed in sections 7.1 and 7.2 are suggested in order to determine the plastic deformation for different orientation distributions:

- Tensile tests with samples extracted from a part with an FOD which significantly differs from that of the parts used in 7.1 should be carried out. The tensile tests should be performed similarly to those suggested in section 7.1. An approximately unidirectional fibre orientation is recommended.

- In case alternative stress states are relevant, the experiments presented in 7.2 also have to be conducted with the samples comprising differing FODs.

In engineering practice, additional experiments with samples comprising alternative FODs may not be available. Such can be the case in pre-design stages. For this reason, the capability of applying the material model proposed in this work in such a context was checked. It was assumed that in such a phase at least one tensile stress-strain curve for a composite with known fibre orientation is available. Two cases were considered in this regard. For the first case, the tension curve for PPGF30$^\square$ under 0° loading was applied. For the second case, the tensile behaviour of the 0° PPGF30$^\square$ sample was used.

The model applied in figure 7.1 was parametrised only by data evaluated from 0° tension of PPGF30$^\square$. The respective material parameters are given in table 7.1. The figure shows the response for various orientation distributions and loading directions. It can be seen that the interpolation assumption (5.86) introduced in section 5.3.9 allows the approximate prediction of the mechanical behaviour under tension for orientations that do not correspond to the configuration that was used for parametrisation.

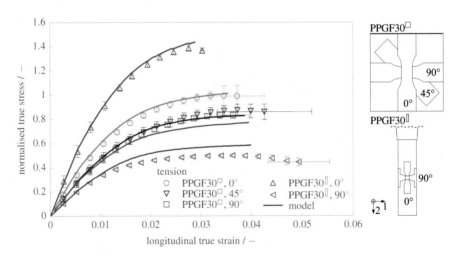

Figure 7.1 Results of the material model compared to tensile experiments. Experimental data from 0° PPGF30$^\square$ was used for parametrisation only.

In figure 7.2, the parametrisation was conducted only from the tensile results of the PPGF30$^\square$ 0° sample. Table 7.1 shows the respective material parameters. The findings are similar to the previous case and the tensile results for alternative orientation distributions can be approximately predicted. This case probably more closely resembles a context that can be found in engineering practice, since in the majority of cases merely results from standard tension bars are available to engineers. These tension bars mostly show a high degree of fibre orientation as is shown in AMBERG, 2004.

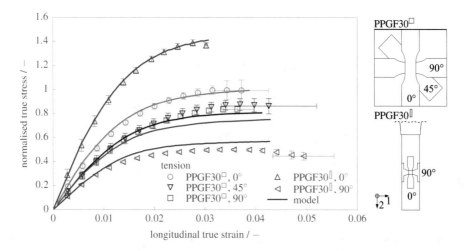

Figure 7.2 Results of the material model compared to tensile experiments. Experimental data from 0° PPGF30▯ was used for parametrisation only.

Table 7.1 Reduced parameter sets for the description of hardening of PPGF30

	Values			
Parameters	Tension from PPGF30▯, 0°		Tension from PPGF30▯, 0°	
$\sigma_{Y,reference}$	23.94 N/mm²		34.83 N/mm²	
ε_{rp}	0.000759		0.000759	
σ_{max} (interpolation)	$\dfrac{\sigma_Y}{\sigma_{Y,reference}}$	$\dfrac{\sigma_{max}}{\sigma_{Y,reference}}$	$\dfrac{\sigma_Y}{\sigma_{Y,reference}}$	$\dfrac{\sigma_{max}}{\sigma_{Y,reference}}$
	0.000	0.000	0.000	0.000
	1.000	3.029	1.000	2.91

The previous examples have shown that the proposed model allows approximative predictions of the tested PPGF30 behaviour even when merely a single stress-strain curve is available for parametrisation. Thus, it is applicable for pre-design phases. This was verified for the tensile stress state. Alternative stress states are approximated from the tensile behaviour as described in section 5.3.9. In order to more adequately predict alternative stress states, experimental data for these stress states should be available.

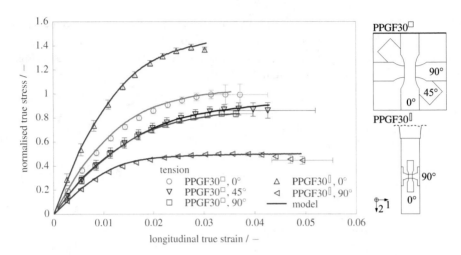

Figure 7.3   Results of model compared to tensile experiments. Data from all of the shown experiments was used for parametrisation.

A benefit of the interpolative aspects of the material model introduced in section 5.3.9 is, that it can be continuously refined with each available experiment. Every experiment that is included in the parametrisation directly leads to a more adequate prediction for the respective orientation distribution. This allows a qualitative improvement of the prediction throughout the entire product development process. When, for example, the four experimental stress-strain curves from figure 7.1 are included for parametrisation, the material model adequately reproduces these curves. This is summarised in figure 7.3.

For the tested PPGF30, it was found that the relation of yield values to maximum stress values varies for different stress states, cf. table 6.7. However, for the case that alternative stress states have not been tested, the interpolation relation applied for the uniaxial tensile stress states considering different FODs can give an approximation of the relations for these stress states.

The parametrisation method presented in section 4.3.2.6 and the normalisation approach discussed in section 4.4.2 can be applied to describe strain rate dependency. If strain rate dependent behaviour is relevant, it is recommended to perform the following additional tests:

- Uniaxial tensile tests with samples extracted in the direction of principal fibre orientation. The tests should be performed with at least three differing testing speeds encompassing the relevant strain rate range.

In reference to section 5.3.7, it is suggested to approximate the plastic Poisson's ratios by the elastic Poisson's ratios computed in section 7.1.

7.4 Consideration of Damage and Failure

If elastic damage shall be taken into account, it is suggested to apply the normalisation approach given in section 4.4.3 and perform the following additional test:

- A uniaxial tensile test with unloading and reloading steps at different stages of deformation is recommended. The experiment should be performed with samples extracted transverse to the principal fibre orientation.

A failure model according to the approach given in section 5.4.2 can be parametrised from the experiments performed in sections 7.1 to 7.3. For the samples extracted from the PPGF30 plate parts in the present work, the failure stresses are at approximately 91 % of the asymptotic stress maximum determined for the inverse polynomial hardening description.

Figure 7.4 presents an overview of the recommendations given in this chapter.

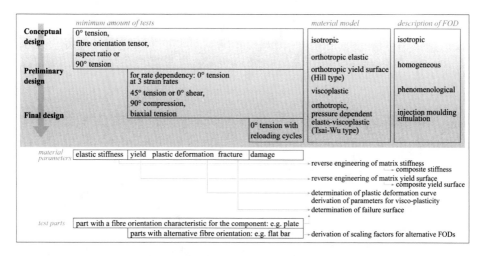

Figure 7.4 Summary of the recommendations for material modelling and parametrisation given in chapter 7. The choice of an appropriate material model depends strongly on the application context.

8 Summary and Outlook

8.1 Summary

The ability to precisely predict mechanical behaviour by FEA is a requirement for the safe design of thermoplastic parts. By proposing, in its essence, a novel material model for explicit finite element simulations of SFRTPs, this work contributes to the realisation of this requirement. The focus was set on the determination of permissible model simplifications based on experimental findings. Hence, this work additionally contributes to the experimental evaluation of SFRTP behaviour.

A PPGF30 composite was specifically compounded for this work, in order to distinctly differentiate the influence of the fibres and the matrix. This allowed for the separate evaluation of matrix properties since the exact composition was known. μCT analysis was used to evaluate the fibre orientation distributions. A novel evaluation approach is presented in order to account for the rotation of the principal axis system during the determination of second order orientation tensor distributions. Mechanical properties were characterised dependent on loading direction, strain rate, stress state, damage history and fibre orientation distributions. In this context, a method for the evaluation of lateral characteristics of uniaxial specimens was proposed. For the assessment of volumetric deformation during tensile tests, a method based on the application of standard 3D DIC was suggested. Regarding the experimental results, central conclusions include that the strain rate dependency of plastic deformation curves is approximately isotropic. Similar results were found for damage curves from cyclic tensile loading. In μCT analysis of preloaded samples, no influence on fibre orientation or average fibre length could be found. A detailed investigation of the deformation of cubic compression samples showed highly inhomogeneous behaviour that can be clearly ascribed to specific fibre distributions.

The proposed material model is based on a coupled micro-mechanical and phenomenological approach. Elastic deformation was assumed to be linear and viscoelastic effects were neglected. The implemented state of the art concept for the computation of the composite stiffness combines a Mori-Tanaka homogenisation scheme for the determination of UD behaviour with an orientation averaging approach. In order to compute the initial yield surface, a novel approach was proposed. A quadratic SAMP yield surface was introduced on the matrix level and related to an orthotropic pressure dependent Tsai-Wu yield surface on the composite level. The hereby derived UD yield surface is expanded to arbitrary orientation distributions by an orientation averaging scheme. It was shown that the parameters of the

F. Dillenberger, *On the anisotropic plastic behaviour of short fibre reinforced thermoplastics and its description by phenomenological material modelling*, Mechanik, Werkstoffe und Konstruktion im Bauwesen 53, https://doi.org/10.1007/978-3-658-28199-1_8

non-associated orthotropic flow potential can be approximated from the elastic Poisson's ratios. Based on the experimental results, a scalar damage formulation was included and strain rate dependency was accounted for by an isotropic scaling of the composite yield surface. To account for fracture a Tsai-Wu limit surface in stress space was implemented.

Isotropic hardening was accounted for phenomenologically by scaling the initial yield surface in dependence of the effective plastic strain. A novel analytical equation based on an inverse polynomial with four parameters was proposed for the description of the hardening curves. The parameters can partly be gained from the initial micro-mechanical computations. The remaining variables depend on the respective fibre orientation. It was assumed that the relation between the initial yield stress and the hardening parameters is unique and can be interpolated. An advantage of the interpolation is that the material parametrisation can continuously be improved with each available experiment. By applying the hardening equation, main parts of the non-linear material behaviour can be determined from a single experiment for each stress state. Moreover, the novel formulation allows for a straightforward parametrisation since the parameter influences can be directly assessed.

The capability of applying the novel material model for pre-design stages in engineering practice was pointed out. It was demonstrated that approximative predictions of the tensile behaviour of the PPGF30 are possible with only a single stress-strain curve available for parametrisation. The findings regarding the novel model can additionally be applied to the parametrisation of existing, already implemented constitutive formulations based on Hill's yield surface. Overall, the numerically implemented constitutive model was verified to be suitable for the prediction of the investigated PPGF30 regarding stiffness, initial yield values and hardening. Tensile tests and compression tests with homogeneous deformation were adequately predicted. In case of shear, the predictions qualitatively resembled the experiments but overestimated the quantitative values occurring at high deformations and multiaxial stress states.

In conclusion, the key achievement of this work is a simplified material model that can predict linear-elastic effects and the non-linear orthotropic pressure dependent viscoplastic deformation of SFRTPs with arbitrary orientation distributions. It was shown that, even if some effects such as the strain rate dependency in the plastic range, occur on the matrix level, a phenomenological approach on the composite level may be assumed. The coupling of the micro-mechanical and phenomenological approach allows the application of complex yield surfaces. Such yield surfaces can only be directly implemented into full micro-mechanical models with a large numerical and computational effort. On the one hand, by coupling both approaches, the computational issues of micro-mechanical models regarding plasticity are avoided. On the other hand, the rather complex parametrisation intrinsic to phenomenological methods can be reduced. A large benefit of the proposed implementation is that all of the micro-mechanical computations for the stiffness and the yield surface are executed only once in the initial time-step. In all post-initial time-steps merely the phenomenological computations need to be processed and no complex micro-mechanical computations are necessary.

8.2 Outlook

In the next step, the material model should be validated on a specimen and a component level in a full FEA framework. Additional efficiency aspects could be considered from works such as e.g. KÖBLER et al., 2018 which propose the clustering of similar orientation distributions.

A verification of the constitutive equations regarding alternative composites with different matrix and fibre materials as well as constituent fractions should be performed. E.g. the range of validity of the isotropic strain rate dependency assumption should be further inspected. SCHÖPFER, 2011 suspected that such an effect is also found in PAGF30. As discussed in section 5.3.10, the micro-mechanical equations for the initial yield stress underline this effect for all SFRTPs if viscoelasticity is neglected. The verification tests have shown that the model should be extended to viscoelasticity when a large range of strain rates is inspected. The analytical hardening equation given in section 5.3.8 will probably need to be adjusted for composites with a higher plastic deformation capability. It remains to be shown if the approximative linear relation found for the asymptotic maximum stress values in section 5.3.9 is valid for alternative SFRTPs.

For the detailed characterisation of SFRTPs over a large range of coupled stress states, HEGLER, 1987, PFLAMM-JONAS, 2001, KAISER, 2013 and others propose tube specimens. Alternatives to the implemented yield surfaces and especially more sophisticated failure approaches should be evaluated. Due to the structure of the material model, these formulations could be straightforwardly implemented. Considering, for example, the determination of the initial yield surface, the orientation averaging approach in the second homogenisation step in section 5.3.1 could be substituted by a weakest link approach. Regarding arbitrarily oriented composites as a combination of UD pseudo grains, global yielding is considered in such an approach as soon as the yield value of any pseudo grain is reached. Alternatively, the anisotropic strength criterion could be formulated by applying the concept of fabric tensors, cf. COWIN, 1986. The current isotropic damage model considers singular loading and unloading processes. It should be expanded for more complex applications with multiple loading cycles and intricate loading paths.

Finally, effects such as temperature and media dependency, as well as ageing strongly influence the SFRTP characteristics and should be included in the constitutive equations. It is conceivable that similar simplifications as have been implemented regarding viscoplastic effects can be applied in this context.

References

ARIF, M. F. (2014): Damage mechanisms in short glass fiber reinforced polyamide-66 under monotic and fatigue loading: Effect of relative humidity and injection molding induced microstructure, PhD thesis, France: Ecole nationale supérieure d'arts et métiers - ENSAM.

ADAM, L., DEPOUHON, A., and ASSAKER, R. (2009): Multi-Scale Modeling of Crash & Failure of Reinforced Plastics Parts with DIGIMAT to LS-DYNA interface, in: *Proceedings of the 7th European LS-Dyna Conference.*

ADVANI, S. G. and TUCKER, C. L. (1985): A tensor description of fiber orientation in short fiber composites, in: *SPE Technical Papers* 13, pp. 1113–1118.

ADVANI, S. G. and TUCKER, C. L. (1987): The use of tensors to describe and predict fiber orientation in short fiber composites, in: *Journal of Rheology* vol. 31.8, pp. 751–784.

AGBOOLA, B. O., JACK, A. D., and MONTGOMERY-SMITH, S. (2012): Effectiveness of recent fiber-interaction diffusion models for orientation and the part stiffness predictions in injection molded short-fiber reinforced composites, in: *Composites Part A: Applied Science and Manufacturing* vol. 43, pp. 1959–1970.

ALTENBACH, H. (2015): *Kontinuumsmechanik: Einführung in die materialunabhängigen und materialabhängigen Gleichungen*, Springer Vieweg.

AMBERG, J. (2004): *Ermittlung temperaturabhängiger anisotroper Stoffwerte für die Spritzgießsimulation*, AiF Research Report 13220 N, Deutsches Kunststoff-Institut DKI.

AMBERG, J. and DILLENBERGER, F. (2013): *Modellierung der orientierungsabhängigen mechanischen Eigenschaften von kohlenstofffaserverstärkten Thermoplastformteilen*, AiF Research Report 16712 N, Fraunhofer Institute for Structural Durability and System Reliability LBF.

AZZI, V. D. and TSAI, S. W. (1965): Anisotropic strength of composites, in: *Experimental Mechanics* vol. 5.9.

BADER, M. G., BAILEY, J. E., and BELL, I. (1973): The effect of fibre-matrix interface strength on the impact and fracture properties of carbon-fibre-reinforced epoxy resin composites, in: *Journal of Physics D: Applied Physics* vol. 6.5, pp. 572–583.

BANABIC, D. (2010): *Sheet Metal Forming Processes - Consitutive Modelling and Numerical Simulation*, Springer-Verlag Berlin-Heidelberg.

BARGEL, H.-J. and SCHULZE, G. (2008): *Werkstoffkunde*, Springer-Verlag Berlin-Heidelberg.

© Springer Fachmedien Wiesbaden GmbH, part of Springer Nature 2020
F. Dillenberger, *On the anisotropic plastic behaviour of short fibre reinforced thermoplastics and its description by phenomenological material modelling*, Mechanik, Werkstoffe und Konstruktion im Bauwesen 53, https://doi.org/10.1007/978-3-658-28199-1

BAY, R. S. and TUCKER, C. L. I. (1992): Stereological Measurement and Error Estimates for Three-Dimensional Fiber Orientation, in: *Polymer Engineering and Science* vol. 32.4, pp. 240–253.

BECKER, F. (2009): Entwicklung einer Beschreibungsmethodik für das mechanische Verhalten unverstärkter Thermoplaste bei hohen Deformationsgeschwindigkeiten, PhD thesis, Germany: Martin-Luther-University Halle-Wittenberg.

BECKER, F. and KRAATZ, A. (2008): Determination of the Behaviour of Thermoplastics at High Strain-Rates Using the Invariant Theory, in: *Proceedings of the 7th LS-DYNA Forum.*

BENVENISTE, Y. (1987): A new approach to the application of Mori-Tanaka's theory in composite materials, in: *Mechanics of Materials* vol. 6.2, pp. 147 –157.

BERG, C. A. (1972): Construction of the Equivalent Plastic Strain Increment, in: *Studies in Applied Mathematics* vol. 51.3, pp. 311–316.

BERNASCONI, A. and COSMI, F. (2011): Analysis of the dependence of the tensile behaviour of a short fibre reinforced polyamide upon fibre volume fraction, length and orientation, in: *Procedia Engineering* vol. 10, 11th International Conference on the Mechanical Behavior of Materials (ICM11), pp. 2129 –2134.

BERNASCONI, A., COSMI, F., and HINE, P. (2012): Analysis of fibre orientation distribution in short fibre reinforced polymers: A comparison between optical and tomographic methods, in: *Composites Science and Technology* vol. 72.16, pp. 2002 –2008.

BETTEN, J. (1979): Über die Konvexität von Fließkörpern isotroper und anisotroper Stoffe, in: *Acta Mechanica* vol. 32.4, pp. 233–247.

BÖHM, H. J. (2017): *A short introduction to basic aspects of continuum micromechanics*, ILSB-Report 206, Wien, Austria: Institute of Lightwight Design and Structural Biomechanics.

BONNET, M. (2009): *Kunststoffe in der Ingenieuranwendung*, Vieweg + Teubner, GWV Fachverlage GmbH Wiesbaden.

BORST, R. de et al. (2012): *Non-Linear Finite Element Analysis of Solids and Structures*, John Wiley & Sons, Ltd.

BOSSELER, M. (2017): Beschreibung des orthotrop visko-elastoplastischen Verhaltens langglasfaserverstärkten Polypropylens - Versuchskonzept und FE-Simulation, PhD thesis, Germany: Technische Universität Kaiserslautern.

BRONSHTEIN, I. N. et al. (2007): *Handbook of Mathematics, 5th Edition*, Springer-Verlag Berlin-Heidelberg.

BROWN, N. and WARD, L. M. (1968): Load Drop at the Upper Yield Point of a Polymer, in: *Journal of Polymer Science Part A-2: Polymer Physics* vol. 6, pp. 607–620.

BRYLKA, B. et al. (2018): DMA based characterization of stiffness reduction in long fiber reinforced polypropylene, in: *Polymer Testing* vol. 66, pp. 296 –302.

CAMACHO, C. W. and AL., et (1990): Stiffness and thermal expansion predictions for hybrid short fiber composites, in: *Polymer Composites* vol. 11.4, pp. 229–239.

CHAVES, E. (2013): *Notes on Continuum Mechanics*, Lecture Notes on Numerical Methods in Engineering and Sciences, Springer Netherlands.

CHOW, C. L. and WANG, J. (1987): An anisotropic theory of elasticity for continuum damage mechanics, in: *International Journal of Fracture* vol. 33, pp. 3–16.

CHRISTENSEN, R. (2013): *The Theory of Materials Failure*, Oxford University Press.

CHUNG, D. H. and KWON, T. H. (2001): Improved model of orthotropic closure approximation for flow induced fiber orientation, in: *Polymer Composites* vol. 22, pp. 636–649.

CINTRA, J. S. and TUCKER, L. I. (1995): Orthotropic closure approximations for flow-induced fiber orientation, in: *Journal of Rheology* vol. 39, pp. 1095–1122.

COWIN, S. C. (1986): Fabric dependence of an anisotropic strength criterion, in: *Mechanics of Materials* vol. 5.3, pp. 251–260.

COX, H. L. (1952): The elasticity and strength of paper and other fibrous materials, in: *British Journal of Applied Physics* vol. 3, pp. 72–79.

CUNTZE, R. G. and FREUND, A. (2004): The predicitve capability of failure mode concept-based strength criteria for multidirectional laminates, in: *Composites Science and Technology* vol. 63.3, pp. 343–377.

DEAN, A. et al. (2018): Invariant-Based Finite Strain Anisotropic Material Model for Fiber-Reinforced Composites, in: *Multiscale Modeling of Heterogeneous Structures*, ed. by J. SORIĆ, P. WRIGGERS, and O. ALLIX, Springer International Publishing, pp. 83–110.

DILLENBERGER, F. (2014): *Finite-Element-Validation - A New Method for Finite-Element-Validation in Crashworthiness Analysis using Optical Sensors*, AiF Report 59 EN, Fraunhofer Institute for Structural Durability and System Reliability LBF.

DILLENBERGER, F. (2011): Ermittlung von Druckkennwerten an Kunststoffen, Masters Thesis, Germany: Technical University Darmstadt.

DOGHRI, I. and TINEL, L. (2005): Micromechanical modeling and computation of elasto-plastic materials reinforced with distributed-orientation fibers, in: *International Journal of Plasticity* vol. 21.10, pp. 1919–1940.

DOGHRI, I. and TINEL, L. (2006): Micromechanics of inelastic composites with mis-aligned inclusions: Numerical treatment of orientation, in: *Computer Methods in Applied Mechanics and Engineering* vol. 195.13, pp. 1387–1406.

DÖLL, T. (2016): Modellierung des plastischen Verhaltens thermoplastischer Materialien in Abhängigkeit von der Spannungstriaxialität, Masters Thesis, Germany: Technical University Darmstadt.

DOMININGHAUS, H. (2008): *Kunststoffe*, Springer-Verlag Berlin-Heidelberg.

DRASS, M., SCHNEIDER, J., and KOLLING, S. (2018): Damage effects of adhesives in modern glass façades: a micro-mechanically motivated volumetric damage model for poro-hyperelastic materials, in: *International Journal of Mechanics and Materials in Design* vol. 14.4, pp. 591–616.

DRAY, D., REGNIER, G., and GILORMINI, P. (2007): Comparison of several closure approximations for evaluating the thermoelastic properties of an injection molded short-fiber composite, in: *Composites Science and Technology* vol. 67, pp. 1601–1610.

DUNNE, F. and PETRINIC, N. (2006): *Introduction to Computational Plasticity*, Oxford University Press New York.

DVORAK, G. J. and BAHEI-EL-DIN, Y. A. (1987): A bimodal plasticity theory of fibrous composite materials, in: *Acta Mechanica* vol. 69.1, pp. 219–241.

EIK, M. (2014): Orientation of short steel fibres in concrete: measuring and modelling, PhD thesis, Finland: Aalto University.

ESHELBY, J. D. (1957): The determination of the elastic field of an ellipsoidal inclusion, and related problems, in: *Proceedings of the Royal Society of London* vol. 241.A, pp. 376–396.

EYERER, P. and AL., et (2008): *Polymer Engineering - Technologien und Praxis*, Springer-Verlag Berlin-Heidelberg.

FENG, W. and YANG, W. (1984): General and Specific Quadratic Yield Functions, in: *Journal of Composites, Technology and Research* vol. 6.1, pp. 19–21.

FISCHER, G. and EYERER, P. (1988): Measuring spatial orientation of short fiber reinforced thermoplastics by image analysis, in: *Polymer Composites* vol. 9.4, pp. 297–304.

FLIEGENER, S., KENNERKNECHT, T., and KABEL, M. (2017): Investigations into the damage mechanisms of glass fiber reinforced polypropylene based on micro specimens and precise models of their microstructure, in: *Composites Part B: Engineering* vol. 112, Supplement C, pp. 327 –343.

FORNOFF, M. and DILLENBERGER, F. (2018): *Phänomenologische Berechnungsstrategie für kurzfaserverstärkte Spritzgussformteile*, AiF Report 16712 N, Fraunhofer Institute for Structural Durability and System Reliability LBF.

FU, S.-Y et al. (2000): Tensile properties of short-glass-fiber- and short-carbon-fiber-reinforced polypropylene composites, in: *Composites Part A: Applied Science and Manufacturing* vol. 31.10, pp. 1117 –1125.

FU, S.-Y., LAUKE, B., and MAI, Y.-M. (2009): *Science and engineering of short fibre reinforced polymer composites*, Woodhead Publishing Limited Cambridge.

FU, S.-Y. and LAUKE, B. (1996): Effects of fiber length and fiber orientation distributions on the tensile strength of short-fiber-reinforced polymers, in: *Composites Science and Technology* vol. 56, pp. 1179–1190.

FUCHS, P. (2007): Anwendung von neuronalen Netzen für die Materialdatengenerierung am Beispiel von Polyamid, Diplomas Thesis, Austria: Montanuniversität Leoben.

GAVAZZI, A. C. and LAGOUDAS, D. C. (1990): On the numerical evaluation of Eshelby's tensor and its application to elastoplastic fibrous composites, in: *Computational Mechanics* vol. 7, pp. 13–19.

GEVATTER, H.-J. and GRÜNHAUPT, U. (2006): *Handbuch der Mess- und Automatisierungstechnik in der Produktion*, Springer-Verlag Berlin-Heidelberg.

GLASER, S., WÜST, A., and AUMER, B. (2008): Metall ist die virtuelle Messlatte, in: *Kunststoffe* vol. 7, pp. 88–92.

GLÖCKNER, R. (2013): *Schnelle quantitative Faserstrukturanalyse mittels Computertomographie*, AiF Research Report 16393 N, Fraunhofer Institute for Structural Durability and System Reliability LBF.

GLÖCKNER, R., KOLLING, S., and HEILIGER, C. (2016): A Monte-Carlo Algorithm for 3D Fibre Detection from Microcomputer Tomography, in: *Journal of Computational Engineering* vol. 2016, pp. 1–9.

GOL'DENBLAT, I. I. and KOPNOV, V. A. (1965): Strength of glass-reinforced plastics in the complex stress state, in: *Polymer Mechanics* vol. 1.2, pp. 54–59.

GROSS, D. and SEELIG, T. (2011): *Bruchmechanik mit einer Einführung in die Mikromechanik*, Springer-Verlag Berlin-Heidelberg.

GRUBER, G., MALTZ, N., and WARTZACK, S. (2011): Berücksichtigung anisotroper Materialeigenschaften crashbelasteter Leichtbaustrukturen im Kontext früher Entwicklungsphasen, in: *Design for X. Proceedings of the 22. DfX-Symposium, Tutzing*, ed. by D. KRAUSE, K. PAETZOLD, and S. WARTZACK, pp. 217–228.

GRUBER, G., HAIMERL, A., and WARTZACK, S. (2012): Consideration of Orientation Properties of Short Fiber Reinforced Polymers within Early Design Steps, in: *Proceedings of the 12th International LS-DYNA Users Conference*.

G'SELL, C. and JONAS, J. J. (1979): Determination of the plastic behaviour of solid polymers at constant true strain rate, in: *Journal of Materials Science* vol. 14.3, pp. 583–591.

GÜNZEL, S. (2013): Analyse der Schädigungsprozesse in einem kurzglasfaserverstärkten Polyamid unter mechanischer Belastung mittels Röntgenrefraktometrie, Bruchmechanik und Fraktografie, PhD thesis, Germany: Technische Universität Berlin.

GUSEV, A. A. (1997): Representative volume element size for elastic composites: A numerical study, in: *Journal of Mechanics and Physics of Solids* vol. 45, pp. 1449–1459.

HABERSTROH, E. and BISTER, H. (2006): Vorhersage der Faserorientierung und der mechanischen Eigenschaften kurzfaserverstärkter PUR-Bauteile, in: *Journal of Plastics Technology* vol. 2, pp. 1–21.

HALPIN, J. C. and KARDOS, J. L. (1976): The Halpin-Tsai equations: A review, in: *Polymer Engineering and Science* vol. 16, pp. 344–352.

HASHIN, Z. and ROSEN, B. W. (1964): The elastic moduli of fiber-reinforced materials, in: *Journal of Applied Mechanics* vol. 31, pp. 223–232.

HAUFE, A. et al. (2005): A semi-analytical model for polymers subjected to high strain rates, in: *Proceedings of the 5th European LS-DYNA Users Conference*.

HAUFE, A. et al. (2010): Recent Developments in Damage and Failure Modeling with LS-DYNA, in: *Proceedings of the 2010 Nordic LS-DYNA Users Forum*.

HEGLER, R. P. (1984): Faserorientierung beim Verarbeiten kurzfaserverstärkter Thermoplaste, in: *Kunststoffe* vol. 74.5, pp. 271–277.

HEGLER, R. P. (1987): Struktur und mechanische Eigenschaften glaspartikelgefüllter Thermoplaste, PhD thesis, Germany: Technische Hochschule Darmstadt.

HENCKY, H. (1928): Über die Form des Elastizitätsgesetzes bei ideal elastischen Stoffen, in: *Z. Techn. Phys.* Vol. 9, pp. 215–220.

HESSMAN, P. A. and HORNBERGER, K. (2018): A comparison of mean field homogenization schemes for short fiber reinforced thermoplastics, in: *PAMM* vol. 17.1, pp. 595–596.

HILL, R. (1948): A theory of the yielding and plastic flow of anisotropic metals, in: *Proceedings of the Royal Society of London A: Mathematical, Physical and Engineering Sciences* vol. 193.1033, pp. 281–297.

HILL, R. (1963): Elastic properties of reinforced solids: Some theoretical principles, in: *Journal of Mechanics and Physics of Solids* vol. 11, pp. 357–372.

HILL, R. (1965a): A self-consistent mechanics of composite materials, in: *Journal of Mechanics and Physics of Solids* vol. 13, pp. 213–222.

HILL, R. (1965b): Theory of mechanical properties of fibre-strengthened materials. Self-consistent model, in: *Journal of Mechanics and Physics of Solids* vol. 13, pp. 189–198.

HINE, P., LUSTI, H. R., and GUSEV, A. A. (2002): Numerical simulation of the effects of volume fraction, aspect ratio and fibre length distribution on the elastic and thermoelastic properties of short fibre composites, in: *Composites Science and Technology* vol. 62, pp. 1445–1453.

HOFFARTH, C. (2016): A Generalized Orthotropic Elasto-plastic Material Model for Impact Analysis, PhD thesis, USA: Arizona State University.

HOFFARTH, C. et al. (2017): Verification and Validation of a Three-Dimensional Orthotropic Plasticity Constitutive Model Using a Unidirectional Composite, in: *Fibers* vol. 5.12, pp. 1–13.

JENKIN, C. F. (1920): *Report on Materials Used in the Construction of Aircraft and Aircraft Engines*, Aeronautical Research Committee, His Majesty's Stationery Office, London.

JENNRICH, R. et al. (2014): Experimental and Numerical Analysis of a Glass Fiber Reinforced Plastic, in: *Proceedings of the 13th LS-DYNA Forum*.

JERABEK, M., MAJOR, Z., and LANG, R. W. (2010): Uniaxial compression testing of polymeric materials, in: *Polymer Testing* vol. 29, pp. 302–309.

JUNGINGER, M. (2002): Charakterisierung und Modellierung unverstärkter thermoplastischer Kunststoffe zur numerischen Simulation von Crashvorgängen, PhD thesis, Germany: Universität der Bundeswehr, Munich.

KADDOUR, A. S. and HINTON, M. J. (2013): Maturity of 3D failure criteria for fibre-reinforced composites: Comparison between theories and experiments: Part B of WWFE-II, in: *Journal of Composite Materials* vol. 47.6-7, pp. 925–966.

KAISER, J.-M. (2013): Beitrag zur mikromechanischen Berechnung kurzfaserverstärkter Kunststoffe - Deformation und Versagen, PhD thesis, Germany: Saarland University.

KAISER, J.-M. and STOMMEL, M. (2014): Modified mean-field formulations for the improved simulation of short fiber reinforced thermoplastics, in: *Composites Science and Technology* vol. 99, pp. 75–81.

KAMMOUN, S. et al. (2011): Micromechanical modeling of short glass-fiber reinforced thermoplastics - Isotropic damage of pseudograins, in: *AIP Conference Proceedings* vol. 1353.1, pp. 972–977.

KEHRER, L. et al. (2015): Experimental investigation and approximation of the temperature-dependent stiffness of short-fiber reinforced polymers, in: *PAMM* vol. 15.1, pp. 453–454.

KELLY, A. and TYSON, W. (1965): Tensile properties of fibre-reinforced metals: Copper/tungsten and copper/molybdenum, in: *Journal of the Mechanics and Physics of Solids* vol. 13.6, pp. 329 –350.

KLUSEMANN, B. and SVEDSON, B. (2009): Homogenization methods for multi-phase elastic composites: Comparisons and benchmarks, in: *Technische Mechanik* vol. 30.4, pp. 374–386.

KÖBLER, J. et al. (2018): Fiber orientation interpolation for the multiscale analysis of short fiber reinforced composite parts, in: *Computational Mechanics* vol. 61.6, pp. 729–750.

KOLLING, S. et al. (2005): SAMP-1: A Semi-Analytical Model for the Simulation of Polymers, in: *Proceedings of the 4th LS-DYNA Forum* vol. A-II-27, Keynote.

KOLLING, S., BECKER, F., and SCHÖPFER, J. (2010): Material Models of Plastics for Crash Simulation - Research State of the Art, in: *Proceedings of the 2nd Automotive CAE Grand Challenge*.

KOLUPAEV, V. A. (2018): *Equivalent Stress Concept for Limit State Analysis*, Advanced Structured Materials, Springer International Publishing.

KOLUPAEV, V. A. and BOLCHOUN, A. (2008): Kombinierte Fliess - und Grenzbedingungen, in: *Forschung Ingenieurwesen* vol. 72, pp. 209–232.

KOUKAL, A. (2014): Crash- und Bruchverhalten von Kunststoffen im Fußgängerschutz von Fahrzeugen, PhD thesis, Germany: Technical University of Munich.

KRIVACHY, R. (2007): *Charakterisierung und Modellierung kurzfaserverstärkter thermoplastischer Kunststoffe zur numerischen Simulation von Crashvorgängen*, Epsilon - Forschungsergebnisse aus der Kurzzeitdynamik, Fraunhofer-Institut für Kurzzeitdynamik, Ernst-Mach-Institut.

KRIVACHY, R. et al. (2008): Characterisation and modelling of short fibre reinforced polymers for numerical simulation of a crash, in: *International Journal of Crashworthiness* vol. 13.5, pp. 559–566.

KUNKEL, F. (2018): Zum Deformationsverhalten von spritzgegossenen Bauteilen aus talkumgefüllten Thermoplasten unter dynamischer Beanspruchung, PhD thesis, Germany: Otto von Guericke University Magedeburg.

KUYKENDALL, W. P. et al. (2012): On the Existence of Eshelby's Equivalent Ellipsoidal Inclusion Solution, in: *Mathematics and Mechanics of Solids* vol. 17.8, pp. 840–847.

LASPALAS, M. et al. (2007): Application of micromechanical models for elasticity and failure to short fibre reinforced composites. Numerical implementation and experimental validation, in: *Computers and Structures* vol. 86, pp. 977–987.

LEMAITRE, J. and DESMORAT, R. (2005): *Engineering Damage Mechanics: Ductile, Creep, Fatigue and Brittle Failures*, Springer-Verlag Berlin-Heidelberg.

LIN, S. et al. (2013): Stress State Dependent Failure Loci of a Talc-filled Polypropylene Material under Static Loading and Dynamic Loading, in: *Proceedings of the 13th International Conference on Fracture*.

LU, Y. Y. (1998): Computing the logarithm of a symmetric positive definite matrix, in: *Applied Numerical Mathematics* vol. 26.4, pp. 483 –496.

LUSTI, H. R. (2003): Property predictions for short fiber and platelet-filled materials by finite element calculations, PhD thesis, Switzerland: ETH Zurich.

MADDAH, H. A. (2016): Polypropylene as a Promising Plastic: A Review, in: *American Journal of Polymer Science* vol. 6.1, pp. 1–11.

MATZENMILLER, A., LUBLINER, J., and TAYLOR, R. (1995): A constitutive model for anisotropic damage in fiber-composites, in: *Mechanics of Materials* vol. 20.2, pp. 125 –152.

MENYHÁRD, A. et al. (2015): Direct correlation between modulus and the crystalline structure in isotactic polypropylene, in: *Express Polymer Letters* vol. 9.3, pp. 308–320.

MEUWISSEN, M. H. H. (1995): *Yield criteria for anisotropic elasto-plastic metals*, Internal Report WFW 95.152, Faculteit der Werktuigbouwkunde, Technical University Eindhoven.

MICHAELI, E. H. et al. (2007): Integrative Simulation: Vorhersage des Verformungsverhaltens teilkristalliner Thermoplastbauteile mittels einer durchgängigen Simulationskette; in: *Zeitschrift Kunststofftechnik/Journal of Plastics Technology* vol. 3.1, pp. 1–17.

MLEKUSCH, B. (1999): Fibre orientation in short-fibre-reinforced thermoplastics - II. Quantitative measurements by image analysis, in: *Composites Science and Technology* vol. 59.4, pp. 547 –560.

MONEKE, M. (2001): Die Kristallisation von verstärkten Thermoplasten während der schnellen Abkühlung unter Druck, PhD thesis, Germany: Technical University Darmstadt.

MÖNNICH, S. (2015): Entwicklung einer Methodik zur Parameteridentifikation für Orientierungsmodelle in der Spritzgießsimulation, PhD thesis, Germany: Otto von Guericke University Magedeburg.

MONTGOMERY-SMITH, S. et al. (2011): Exact tensor closures for the three-dimensional Jeffery's equation, in: *Journal of Fluid Mechanics* vol. 680, pp. 321–335.

MORI, T. and TANAKA, K. (1973): Average stress in matrix and average elastic energy of materials with misfitting inclusions, in: *Acta Metallurgica* vol. 21, pp. 571–574.

MÜLLER, V. (2015): Micromechanical modeling of short-fiber reinforced composites, PhD thesis, Germany: Karlsruhe Institute of Technology.

MÜLLER, V. et al. (2015): Homogenization of elastic properties of short-fiber reinforced composites based on measured microstructure data, in: *Journal of Composite Materials* vol. 50.3, pp. 297–312.

MURA, T. (1982): *Micromechanics of defects in solids*, The Hague: Martinus Nijhoff Publishers The Hague.

NAHAS, M. N. (1986): Survey of failure and post-failure theories of laminated fiber-reinforced composites, in: *Journal of Composite Technology and Research* vol. 4.8, pp. 138–15.

NCIRI, M. (2017): Constitutive behaviour modelling of short fibre reinforced composites under dynamic loading, PhD thesis, France: Université de Valenciennes et du Hainaut-Cambresis.

NEALE, K. W. and LABOSSIÉRE, P. (1989): Failure of composite laminae under biaxial loading, in: *Polymer Composites* vol. 10.5, pp. 285–292.

NELDER, J. A. (1966): Inverse Polynomials, a Useful Group of Multi-Factor Response Functions, in: *Biometrics* vol. 22.1, pp. 128–141.

NOTTA-CUVIER, D. et al. (2013): An efficient modelling of inelastic composites with misaligned short fibres, in: *International Journal of Solids and Structures* vol. 50.19, pp. 2857 –2871.

NUTINI, M. (2010): Simulation of anisotropy with ls-dyna in glass-reinforced, polypropylene-based components, in: *Proceedings of the 9th LS-DYNA Forum* vol. F-IV, pp. 1–12.

NUTINI, M. and VITALI, M. (2017): Interactive failure criteria for glass fibre reinforced polypropylene: validation on an industrial part, in: *International Journal of Crashworthiness* vol. 0.0, pp. 1–15.

ORTIZ, M. and SIMO, J. C. (1986): An analysis of a new class of integration algorithms for elastoplastic constitutive relations, in: *International Journal for Numerical Methods in Engineering* vol. 23.3, pp. 353–366.

OSSWALD, P. V. and OSSWALD, T. A. (2017): A strength tensor based failure criterion with stress interactions, in: *Polymer Composites* vol. 39.8, pp. 2826–2834.

OSSWALD, T. A. et al. (2006): *International Plastics Handbook*, Carl Hanser Verlag GmbH & Co. KG Munich.

PAETOW, R. (1991): Über das Spannungs-Verformungs-Verhalten von Papier., PhD thesis, Germany: Technische Hochschule Darmstadt.

PALANIVELU, S. et al. (2009): Validation of digital image correlation technique for impact loading applications, in: *Proceedings of DYMAT - International Conference on the Mechanical and Physical Behaviour of Materials under Dynamic Loading* vol. 1, pp. 373–379.

PAPATHANASIOU, T. D. and GUELL, D. C. (1997): *Flow-Induced Alignment in Composite Materials*, Woodhead Publishing Series in Composites Science and Engineering, Elsevier Science.

PEGORETTI, A., ACCORSIS, M. L., and DIBENEDETTO, A. T. (1996): Fracture toughness of the fibre-matrix interface in glass-epoxy composites, in: *Journal of Materials Science* vol. 31.5, pp. 6145–6153.

PERZYNA, P. (1966): Fundamental problems in viscoplasticity, in: *Advances in Applied Mechanics* vol. 9, pp. 243–377.

PETERS, P. R. (2015): Yield Functions taking into account Anisotropic Hardening Effects for an Improved Virtual Representation of Deep Drawing Processes, PhD thesis, Switzerland: ETH Zurich.

PFEIFER, F., KASTNER, J., and FREYTAG, R. (2008): Method for three-dimensional evaluation and visualization of the distribution of fibres in glass-fibre reinforced injection molded parts by X-ray computed tomography, in: *Proceedings of the 17th World Conference on Nondestructive Testing*.

PFEIFFER, M. and KOLLING, S. (2019): A non-associative orthotropic plasticity model for paperboard under in-plane loading, in: *International Journal of Solids and Structures* vol. 166, pp. 112 –123.

PFLAMM-JONAS, T. (2001): Auslegung und Dimensionierung von kurzfaserverstärkten Spritzgussbauteilen, PhD thesis, Germany: Technical University Darmstadt.

PIERARD, O. et al. (2007): Micromechanics of elasto-plastic materials reinforced with ellipsoidal inclusions, in: *International Journal of Solids and Structures* vol. 44, pp. 6945–6962.

PIERARD, O. (2006): Micromechanics of inclusion-reinforced composites in elasto-plasticity and elasto-viscoplasticity: modeling and computation, PhD thesis, Belgium: Université Catholique de Louvain.

PONGRATZ, S. (2000): Alterung von Kunststoffen während der Verarbeitung und im Gebrauch, PhD thesis, Germany: University Erlangen-Nürnberg.

PUCK, A. (1996): *Festigkeitsanalyse von Faser-Matrix-Laminaten: Modelle für die Praxis*, Hanser.

QUI, Y. P. and WENG, G. J. (1990): On the application of Mori-Tanaka's theory involving transversely isotropic spheroidal inclusions, in: *International Journal of Engineering Science* vol. 28.11, pp. 1121–1137.

QUINSON, R. et al. (1997): Yield criteria for amorphous glassy polymers, in: *Journal of Materials Science* vol. 32, pp. 1371–1379.

REGNIER, G., DRAY, D., and GILORMINI, P. (2008): Assessment of the Thermoelastic Properties of an Injection molded short-fiber Composite: Experimental and Modelling, in: *International Journal of Material Forming* vol. 1.1, pp. 787–790.

REITHOFER, P. and FERTSCHEJ, A. (2016): Dynamic material characterization using 4a impetus, in: *AIP Conference Proceedings* vol. 1779, p. 05008.

REITHOFER, P. et al. (2018): *MAT_4a_MICROMEC - Theory and Application Notes, in: *Proceedings of the 15th International LS-DYNA Conference.*

RÖHRIG, C., SCHEFFER, T., and DIEBELS, S. (2017): Mechanical characterization of a short fiber-reinforced polymer at room temperature: experimental setups evaluated by an optical measurement system, in: *Continuum Mechanics and Thermodynamics* vol. 29.5, pp. 1093–1111.

RÖSLER, J., HARDERS, H., and BÄKER, M. (2008): *Mechanisches Verhalten der Werkstoffe*, Vieweg + Teubner, GWV Fachverlage GmbH Wiesbaden.

RÜHL, A. (2017): *On the Time and Temperature Dependent Behaviour of Laminated Amorphous Polymers Subjected to Low-Velocity Impact*, Mechanik, Werkstoffe und Konstruktion im Bauwesen, Springer-Verlag Berlin-Heidelberg.

RUSSEL, W. B. (1973): On the effective moduli of composite materials: Effect of fiber length and geometry at dilute concentrations, in: *Zeitschrift für angewandte Mathematik und Physik ZAMP* vol. 24.4, pp. 581–600.

SAITO, H. et al. (2017): Direct three-dimensional imaging of the fracture of fiber-reinforced plastic under uniaxial extension: Effect of adhesion between fibers and matrix, in: *Polymer* vol. 116, Supplement C, pp. 556 –564.

SAMBALE, A. K., SCHÖNEICH, M., and STOMMEL, M. (2017): Influence of the Processing Parameters on the Fiber-Matrix-Interphase in Short Glass Fiber-Reinforced Thermoplastics, in: *Polymers* vol. 9.6, pp. 1–23.

SAWADA, T. and KUSAKA, T. (2017): Strength predictions by applied effective volume theory in short-glass-fibre-reinforced plastics, in: *Polymer Testing* vol. 62, Supplement C, pp. 143 –153.

SCHAAF, A. (2015): Short fiber reinforced thermoplastics under thermo-mechanical fatigue loading: experimental and numerical considerations, PhD thesis, Germany: Technical University Darmstadt.

SCHMACHTENBERG, E. M. (1985): Die mechanischen Eigenschaften nichtlinear visko-elastischer Werkstoffe, PhD thesis, Germany: RWTH Aachen University.

SCHÖNEICH, M. (2016): Charakterisierung und Modellierung viskoelastischer Eigenschaften von kurzglasfaserverstärkten Thermoplasten mit Faser-Matrix Interphase, PhD thesis, France: Université de Lorraine.

SCHÖPFER, J. (2011): Spritzgussbauteile aus kurzfaserverstärkten Kunststoffen: Methoden der Charakterisierung und Modellierung zur nichtlinearen Simulation von statischen und crashrelevanten Lastfällen, PhD thesis, Germany: Technische Universität Kaiserslautern.

SCHOSSIG, M. (2010): *Schädigungsmechanismen in faserverstärkten Kunststoffen: Quasistatische und dynamische Untersuchungen*, Vieweg + Teubner research, Vieweg + Teubner Verlag.

SCHOSSIG, M. et al. (2008): Mechancial behaviour of glass-fiber reinforced thermoplastic materials under high strain rates, in: *Polymer Testing* vol. 27, pp. 893–900.

SCHRÖDER, J. and NEFF, P. (2010): *Poly-, Quasi- and Rank-One Convexity in Applied Mechanics*, CISM International Centre for Mechanical Sciences, Springer Vienna.

SCHÜRMANN, H. (2007): *Konstruieren mit Faser-Kunststoff-Verbunden*, Springer-Verlag Berlin-Heidelberg.

SHEN, H., NUTT, S., and HULL, D. (2004): Direct observation and measurement of fiber architecture in short fiber-polymer composite foam through micro-CT imaging, in: *Composites Science and Technology* vol. 64.13, pp. 2113 –2120.

SIMO, J. and HUGHES, T. (1998): *Computational Inelasticity*, Springer.

SKRZYPEK, J. and GANCZARSKI, A. (2016): Constraints on the applicability range of pressure-sensitive yield/failure criteria: strong orthotropy or transverse isotropy, in: *Acta Mechanica* vol. 227.8, pp. 2275–2304.

SOUZA NETO, E. de, PERIC, D., and OWEN, D. (2011): *Computational Methods for Plasticity: Theory and Applications*, Wiley.

STEINKE, P. (2007): *Finite-Elemente-Methode*, Springer-Verlag Berlin-Heidelberg.

STOMMEL, M., STOJEK, M., and KORTE, W. (2018): *FEM zur Berechnung von Kunststoff- und Elastomerbauteilen*, Carl Hanser Verlag Munich.

SUQUET, P. (2014): *Continuum Micromechanics*, CISM International Centre for Mechanical Sciences, Springer Vienna.

TANDON, G. P. and WENG, G. J. (1984): The effect of aspect ratio of inclusions on the elastic properties of unidirectionally aligned composites, in: *Polymer Composites* vol. 5, pp. 327–333.

TANG, P. (1979): *A recommendation of a triaxial failure theory for graphite*, Research Report, General Atomic Company and United States, Department of Energy, San Francisco Operations Office.

TANG, P. (1989): *A Multiaxial Failure Criterion for composites*, Research Report, AVAL OCEAN SYSTEMS CENTER SAN DIEGO CA.

TING, T. (1996): *Anisotropic Elasticity: Theory and Applications*, Oxford Engineering Science Series, Oxford University Press.

TSAI, S. W. (1965): *Strength characteristics of composite materials*, Research Report, NASA CR-224.

TSAI, S. W. and WU, E. M. (1971): A General Theory of Strength for Anisotropic Materials, in: *Journal of Composite Materials* vol. 5, pp. 58–80.

TUCKER, C. L. and LIANG, E. (1999): Stiffness predictions for unidirectional short-fiber composites: Review and evaluation, in: *Composites Science and Technology* vol. 59, pp. 655–671.

VAN HATTUM, F. W. J. and BERNARDO, C. A. (1999): A model to predict the strength of short fiber composites, in: *Polymer Composites* vol. 20.4, pp. 524–533.

VOGLER, M. (2005): Implementierung von visko-plastischen Stoffgesetzen für Thermoplaste in LS-DYNA, Masters Thesis, Germany: Brandenburg University of Technology Cottbus.

VOGLER, M. (2014): Anisotropic Material Models for Fiber Reinforced Polymers, PhD thesis, Germany: Leibniz University Hannover.

VOYIADJIS, G. and THIAGARAJAN, G. (1995): An anisotropic yield surface model for directionally reinforced metal-matrix composites, in: *International Journal of Plasticity* vol. 11.8, pp. 867 –894.

WANG, W. M., SLUYS, L. J., and BORST, R. de (1997): Viscoplasticity for instabilities due to strain softening and strain-rate softening, in: *International Journal for Numerical Methods in Engineering* vol. 40.20, pp. 3839–3864.

WITHERS, P. J. (1989): The determination of the elastic field of an ellipsoidal inclusion in a transversely isotropic medium, and its relevance to composite materials, in: *Philosophical Magazine A* vol. 59.4, pp. 759–781.

WITZGALL, C. and WARTZACK, S. (2015): Validierung eines Ansatzes zur Simulation kurzfaserverstärkter Thermoplaste in frühen Entwurfsphasen, in: *Design for X. Proceedings of the 26. DfX-Symposium, Herrsching*, ed. by W. S. KRAUSE D. Paetzold K., pp. 63–74.

ZHENG, R., TANNER, R., and FAN, X. (2011): *Injection Molding: Integration of Theory and Modeling Methods*, Springer-Verlag Berlin-Heidelberg.

ZIENKIEWICZ, O., TAYLOR, R., and ZHU, J. (2005): *The Finite Elemente Method: Its Basis and Fundamentals*, Elsevier Butterworth-Heinemann.

Standards

DIN EN ISO 178 (2013): *Kunststoffe - Bestimmung der Biegeeigenschaften*, Standard, Berlin: DIN Deutsches Institut für Normung e.V.

DIN EN ISO 527 (1996): *Kunststoffe - Bestimmung der Zugeigenschaften*, Standard, Berlin: DIN Deutsches Institut für Normung e.V.

DIN EN ISO 604 (2003): *Kunststoffe - Bestimmung von Druckeigenschaften*, Standard, Berlin: DIN Deutsches Institut für Normung e.V.

Manuals

CONVERSE (2018): *Converse Plastics Simulation - Converse Documentation V 4.0*, Software Manual, Bergish-Gladbach, Germany: PART Engineering GmbH.

DIGIMAT (2006): *DIGIMAT. A software for the linear and nonlinear multi-scale modeling of heterogeneous materials*, Software Manual, Louvain-la-Neuve, Belgium: e-Xstream engineering SA.

HALLQUIST, J. (2006): *LS-DYNA Theory Manual*, Software Manual, Livermore, California, USA: Livermore Software Technology Corporation (LSTC).

LS-DYNA (2017): *LS-DYNA Keyword User's Manual, Volume I, LS-DYNA R10.0*, Software Manual, Livermore, California, USA: Livermore Software Technology Corporation (LSTC).

Appendix A

Overview of Experiments and Sample Geometries

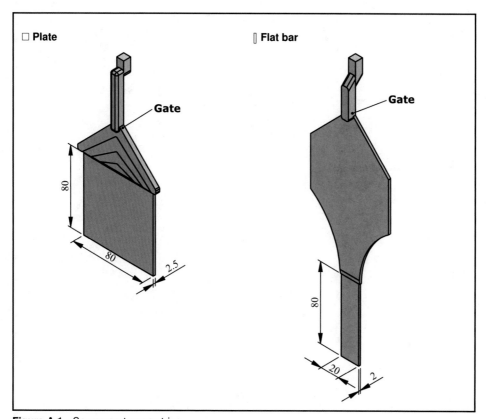

Figure A.1 Source part geometries.

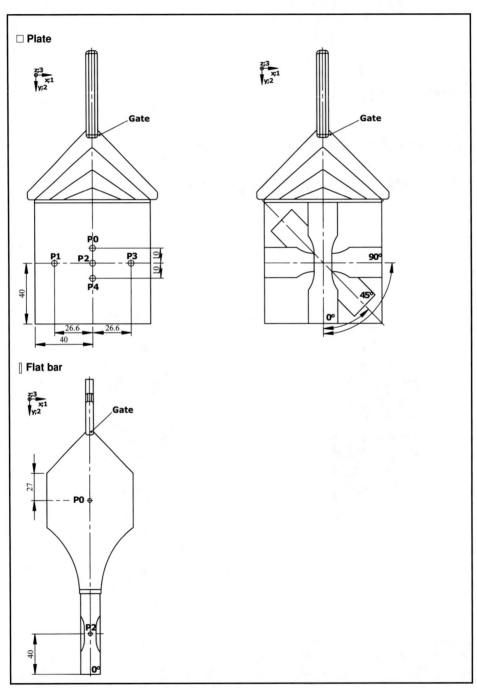

Figure A.2   Specimen extraction positions.

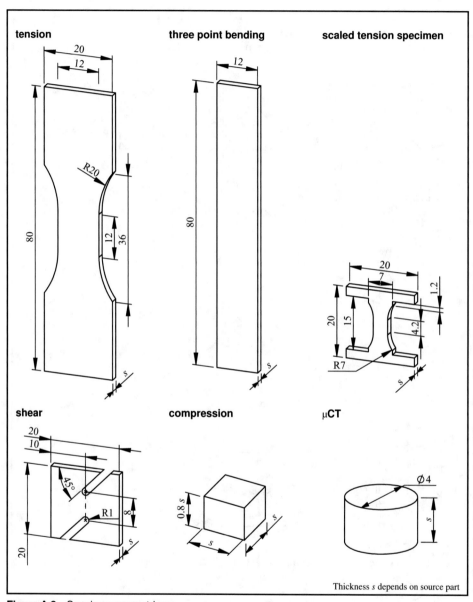

Figure A.3 Specimen geometries.

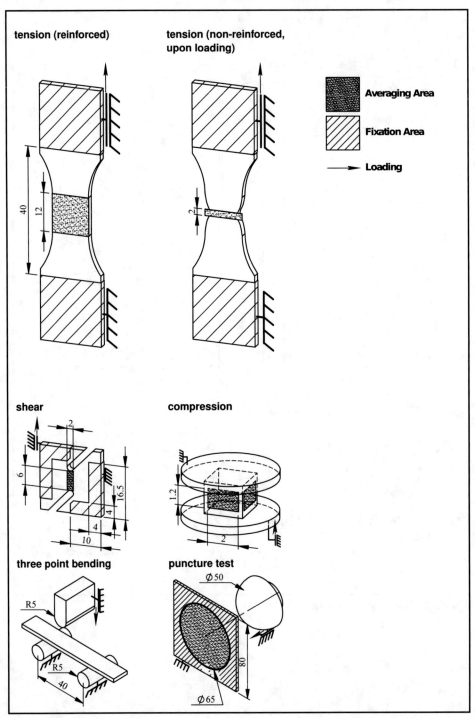

Figure A.4   Experimental set-up.

Appendix B

Equations

B.1 Components of the Eshelby Tensor

Components of the fourth order Eshelby's tensor $\mathbf{L}$ are listed below. The expressions have been extracted from MURA, 1982. Inclusion geometries are defined by the aspect ratio φ. The elastic properties of the isotropic matrix are given by Poisson's ratio v_{12}^m.

For a thin disk with $\varphi \equiv 0$ the following relations can be found:

$$
\begin{aligned}
L_{1111} &= 1, \\
L_{1122} = L_{1133} &= \frac{v_{12}^m}{1 - v_{12}^m}, \\
L_{1212} = L_{1313} &= 1/2.
\end{aligned}
\tag{B.1}
$$

A spherical inclusion is defined by $\varphi = 1$ and leads to the expressions

$$
\begin{aligned}
L_{1111} = L_{2222} = L_{3333} &= \frac{7 - 5v_{12}^m}{15\left(1 - v_{12}^m\right)}, \\
L_{1122} &= \frac{5v_{12}^m - 1}{15\left(1 - v_{12}^m\right)}, \\
L_{2233} = L_{3311} = L_{1133} = L_{2211} = L_{3322} &= L_{1122}, \\
L_{1212} = L_{2323} = L_{3131} &= \frac{4 - 5v_{12}^m}{15\left(1 - v_{12}^m\right)}.
\end{aligned}
\tag{B.2}
$$

For the derivation of spheroidal inclusions an auxiliary variable g is introduced. For oblate spheroidal inclusions with $0 < \varphi < 1$ it is defined as

$$
g = \frac{\varphi}{\left(1 - \varphi^2\right)^{3/2}}\left(\arccos(\varphi) - \varphi\sqrt{1 - \varphi^2}\right).
\tag{B.3}
$$

© Springer Fachmedien Wiesbaden GmbH, part of Springer Nature 2020
F. Dillenberger, *On the anisotropic plastic behaviour of short fibre reinforced thermoplastics and its description by phenomenological material modelling*, Mechanik, Werkstoffe und Konstruktion im Bauwesen 53, https://doi.org/10.1007/978-3-658-28199-1

For prolate spheroidal inclusions with $1 < \varphi < \infty$ it amounts to

$$g = \frac{\varphi}{(\varphi^2 - 1)^{3/2}} \left(\varphi \sqrt{\varphi^2 - 1} - \ln \left(\varphi + \sqrt{\varphi^2 - 1} \right) \right). \tag{B.4}$$

Eshelby's tensor components for spheroidal inclusions can now be given as

$$L_{1111} = \frac{1}{2(1 - v_{12}^m)} \left(1 - 2v_{12}^m + \frac{3\varphi^2 - 1}{\varphi^2 - 1} - \left(1 - 2v_{12}^m + \frac{3\varphi^2}{\varphi^2 - 1} \right) g \right),$$

$$L_{2222} = L_{3333} = \frac{3}{8(1 - v_{12}^m)} \frac{\varphi^2}{\varphi^2 - 1} + \frac{1}{4(1 - v_{12}^m)} \left(1 - 2v_{12}^m - \frac{9}{4(\varphi^2 - 1)} \right) g,$$

$$L_{2233} = L_{3322} = \frac{1}{4(1 - v_{12}^m)} \left(\frac{\varphi^2}{2(\varphi^2 - 1)} - \left(1 - 2v_{12}^m + \frac{3}{\varphi^2 - 1.} \right) g \right),$$

$$L_{2211} = L_{3311} = \frac{-1}{2(1 - v_{12}^m)} \frac{\varphi^2}{\varphi^2 - 1} + \frac{1}{4(1 - v_{12}^m)} \left(\frac{3\varphi^2}{\varphi^2 - 1} - (1 - 2v_{12}^m) \right) g,$$

$$L_{1122} = L_{1133} = \frac{-1}{2(1 - v_{12}^m)} \left(1 - 2v_{12}^m + \frac{1}{\varphi^2 - 1} \right)$$
$$+ \frac{1}{2(1 - v_{12}^m)} \left(1 - 2v_{12}^m + \frac{3}{2(\varphi^2 - 1)} \right) g,$$

$$L_{2323} = \frac{1}{4(1 - v_{12}^m)} \left(\frac{\varphi^2}{2(\varphi^2 - 1)} + \left(1 - 2v_{12}^m - \frac{3}{4(\varphi^2 - 1)} \right) g \right),$$

$$L_{1212} = L_{1313} = \frac{1}{4(1 - v_{12}^m)} \left(1 - 2v_{12}^m - \frac{\varphi^2 + 1}{\varphi^2 - 1} - \frac{1}{2} \left(1 - 2v_{12}^m - \frac{3(\varphi^2 + 1)}{(\varphi^2 - 1)} \right) g \right). \tag{B.5}$$

Finally, for cylindrical inclusions that are defined by $\varphi \to \infty$ the components of Eshelby's tensor result in

$$L_{1111} = L_{1122} = L_{1133} = 0,$$

$$L_{2222} = L_{3333} = \frac{5 - 4v_{12}^m}{8(1 - v_{12}^m)},$$

$$L_{2233} = L_{3322} = \frac{4v_{12}^m - 1}{8(1 - v_{12}^m)},$$

$$L_{2211} = L_{3311} = \frac{v_{12}^m}{2(1 - v_{12}^m)}, \tag{B.6}$$

$$L_{2323} = \frac{3 - 4v_{12}^m}{8(1 - v_{12}^m)},$$

$$L_{1212} = L_{1313} = 1/4.$$

Additional terms are determined by taking into account the symmetries of Eshelby's tensor given by

$$L_{jikl} = L_{ijkl} \quad \text{and} \quad L_{ijlk} = L_{ijkl}. \tag{B.7}$$

Applying these symmetries, the fourth order Eshelby's tensor can be reduced to a second order matrix by means of Voigt's notation. KUYKENDALL et al., 2012 suggest the matrix form

$$L = \begin{bmatrix} L_{1111} & L_{1122} & L_{1133} & L_{1123} & L_{1113} & L_{1112} \\ L_{2211} & L_{2222} & L_{2233} & L_{2223} & L_{2213} & L_{2212} \\ L_{3311} & L_{3322} & L_{3333} & L_{3323} & L_{3313} & L_{3312} \\ 2L_{2311} & 2L_{2322} & 2L_{2333} & 2L_{2323} & 2L_{2313} & 2L_{2312} \\ 2L_{1311} & 2L_{1322} & 2L_{1333} & 2L_{1323} & 2L_{1313} & 2L_{1312} \\ 2L_{1211} & 2L_{1222} & 2L_{1233} & 2L_{1223} & 2L_{1213} & 2L_{1212} \end{bmatrix}. \tag{B.8}$$

B.2 Explicit Formulae for the Effective Stiffness Tensor of the Mori-Tanaka Model

In the following, an explicit formulation for the micro-mechanical Mori-Tanaka model is presented. The expressions allow for an approximation of the effective stiffness tensor of a unidirectional composite consisting of an isotropic matrix and transversely isotropic fibres. The expressions given are based on a simplification of the equations originally derived by QUI et al., 1990 and account for the inclusion of a single fibre material.

As presented in section 3.3, five independent variables are sufficient to define the stiffness tensor of a transversely isotropic material. Following the approach of QUI et al., 1990, the stiffness tensor C^i in Voigt's notation for a material i is defined by the five independent terms χ_i^i, with $i = 0, 1, \ldots 4$:

$$C^i = \begin{bmatrix} \chi_1^i & \chi_4^i & \chi_4^i & 0 & 0 & 0 \\ \chi_4^i & \chi_0^i + \chi_2^i & \chi_0^i - \chi_2^i & 0 & 0 & 0 \\ \chi_4^i & \chi_0^i - \chi_2^i & \chi_0^i + \chi_2^i & 0 & 0 & 0 \\ 0 & 0 & 0 & \chi_2^i & 0 & 0 \\ 0 & 0 & 0 & 0 & \chi_3^i & 0 \\ 0 & 0 & 0 & 0 & 0 & \chi_3^i \end{bmatrix}. \tag{B.9}$$

For a transversely isotropic fibre material f the stiffness tensor components can be determined from the engineering constants of the fibre by the relations

$$
\begin{aligned}
\chi_1^f &= \frac{1 - \left(\frac{E_{22}^f}{2G_{23}^f} - 1 \right)^2}{\left(E_{22}^f \right)^2 \varsigma_0^f}, \\
\chi_2^f &= G_{23}^f, \\
\chi_3^f &= G_{12}^f, \\
\chi_4^f &= \frac{v_{12}^f \left(\frac{E_{22}^f}{E_{11}^f} \right) \left(\frac{E_{22}^f}{2G_{23}^f} \right)}{\left(E_{22}^f \right)^2 \varsigma_0^f} \quad \text{and} \\
\chi_0^f &= \frac{1 - \left(v_{12}^f \right)^2 \left(\frac{E_{22}^f}{E_{11}^f} \right)}{E_{22}^f E_{11}^f \varsigma_0^f} - \chi_2^f.
\end{aligned}
\tag{B.10}
$$

In this context, the auxiliary variable ς_0^f is given as

$$\varsigma_0^f = \frac{1 - 2\left(v_{12}^f\right)^2 \left(\frac{E_{22}^f}{E_{11}^f}\right) - \left(\frac{E_{22}^f}{2G_{23}^f} - 1\right)^2 - 2\left(v_{12}^f\right)^2 \left(\frac{E_{22}^f}{E_{11}^f}\right)\left(\frac{E_{22}^f}{2G_{23}^f} - 1\right)^2}{E_{11}^f \left(E_{22}^f\right)^2}.$$

(B.11)

In a similar way the isotropic stiffness tensor for the matrix material m is derived as

$$\chi_1^m = \frac{(1 - v^m) E^m}{(1 + v^m)(1 - 2v^m)},$$

$$\chi_2^m = \frac{E^m}{2(1 + v^m)},$$

$$\chi_3^m = \chi_2^m,$$

(B.12)

$$\chi_4^m = \frac{v^m E^m}{(1 + v^m)(1 - 2v^m)} \quad \text{and}$$

$$\chi_0^m = \chi_1^m - \chi_2^m = \chi_4^m + \chi_2^m.$$

The micro-mechanical Mori-Tanaka model is capable of approximating the effective stiffness tensor of a unidirectional composite from the stiffness tensors of the matrix and the fibres, as well as Eshelby's tensor L and the fibre volume fraction ϑ^f, cf. section 5.2.

Applying the Eshelby's tensor L given in Voigt's notation from (B.8), the auxiliary variables ς_i, with $i = 1, 2, \dots 7$, are defined for the subsequent calculations as:

$$\varsigma_1 = 1 + 2\frac{\chi_0^f - (\chi_1^m - \chi_3^m)}{E^m}\left((1 - v^m)(L_{22} + L_{23}) - 2v^m L_{21}\right)$$
$$+ 2\frac{\chi_4^f - \chi_4^m}{E^m}\left((1 - v^m)L_{12} - v^m L_{11}\right),$$

$$\varsigma_2 = 1 + \frac{\chi_1^f - \chi_1^m}{E^m}\left(L_{11} - 2v^m L_{12}\right) + 2\frac{\chi_4^f - \chi_4^m}{E^m}\left((1 - v^m)L_{12} - v^m L_{11}\right),$$

$$\varsigma_5 = 2\frac{\chi_0^f - (\chi_1^m - \chi_3^m)}{E^m}\left((1 - v^m)L_{12} - v^m L_{11}\right) + \frac{\chi_4^f - \chi_4^m}{E^m}\left(L_{11} - 2v^m L_{12}\right),$$

$$\varsigma_6 = \frac{\chi_1^f - \chi_1^m}{E^m}\left(L_{21} - v^m (L_{22} + L_{23})\right)$$
$$+ \frac{\chi_4^f - \chi_4^m}{E^m}\left((1 - v^m)(L_{22} + L_{23}) - 2v^m L_{21}\right),$$

$$\varsigma_3 = 1 + \frac{\chi_2^f - \chi_3^m}{\chi_3^m}L_{44}, \quad \varsigma_4 = 1 + \frac{\chi_3^f - \chi_3^m}{\chi_3^m}L_{66}, \quad \varsigma_7 = \varsigma_1\varsigma_2 - 2\varsigma_5\varsigma_6.$$

(B.13)

Finally, the explicit Mori-Tanaka model equations for the computation of the components of the effective stiffness tensor of the composite material c can be determined as:

$$
\chi_0^c = \frac{\left(-\chi_4^f\varsigma_5\vartheta^f + \chi_0^f\varsigma_2\vartheta^f - \varsigma_7(\chi_1^m - \chi_3^m)(\vartheta^f - 1)\right)\left(\varsigma_1\vartheta^f - \varsigma_7(\vartheta^f - 1)\right)}{\left(\varsigma_1\vartheta^f - \varsigma_7(\vartheta^f - 1)\right)\left(\varsigma_2\vartheta^f - \varsigma_7(\vartheta^f - 1)\right) - 2\varsigma_6\varsigma_5(\vartheta^f)^2}
$$

$$
+ \frac{\left(\chi_4^f\varsigma_1\vartheta^f - 2\chi_0^f\varsigma_6\vartheta^f - \chi_4^m\varsigma_7(\vartheta^f - 1)\right)\varsigma_5\vartheta^f}{\left(\varsigma_1\vartheta^f - \varsigma_7(\vartheta^f - 1)\right)\left(\varsigma_2\vartheta^f - \varsigma_7(\vartheta^f - 1)\right) - 2\varsigma_6\varsigma_5(\vartheta^f)^2},
$$

$$
\chi_4^c = \frac{\left(-\chi_1^f\varsigma_5\vartheta^f + \chi_4^f\varsigma_2\vartheta^f - \chi_4^m\varsigma_7(\vartheta^f - 1)\right)\left(\varsigma_1\vartheta^f - \varsigma_7(\vartheta^f - 1)\right)}{\left(\varsigma_1\vartheta^f - \varsigma_7(\vartheta^f - 1)\right)\left(\varsigma_2\vartheta^f - \varsigma_7(\vartheta^f - 1)\right) - 2\varsigma_6\varsigma_5(\vartheta^f)^2}
$$

$$
+ \frac{\left(\chi_1^f\varsigma_1\vartheta^f - 2\chi_4^f\varsigma_6\vartheta^f - \chi_1^m\varsigma_7(\vartheta^f - 1)\right)\varsigma_5\vartheta^f}{\left(\varsigma_1\vartheta^f - \varsigma_7(\vartheta^f - 1)\right)\left(\varsigma_2\vartheta^f - \varsigma_7(\vartheta^f - 1)\right) - 2\varsigma_6\varsigma_5(\vartheta^f)^2},
$$

$$
\chi_1^c = \frac{\left(-\chi_1^f\varsigma_5\vartheta^f + \chi_4^f\varsigma_2\vartheta^f - \chi_4^m\varsigma_7(\vartheta^f - 1)\right)2\varsigma_6\vartheta^f}{\left(\varsigma_1\vartheta^f - \varsigma_7(\vartheta^f - 1)\right)\left(\varsigma_2\vartheta^f - \varsigma_7(\vartheta^f - 1)\right) - 2\varsigma_6\varsigma_5(\vartheta^f)^2}
$$

$$
+ \frac{\left(\chi_1^f\varsigma_1\vartheta^f - 2\chi_4^f\varsigma_6\vartheta^f - \chi_1^m\varsigma_7(\vartheta^f - 1)\right)\left(\varsigma_2\vartheta^f - \varsigma_7(\vartheta^f - 1)\right)}{\left(\varsigma_1\vartheta^f - \varsigma_7(\vartheta^f - 1)\right)\left(\varsigma_2\vartheta^f - \varsigma_7(\vartheta^f - 1)\right) - 2\varsigma_6\varsigma_5(\vartheta^f)^2},
$$

$$
\chi_2^c = \frac{\chi_2^f\vartheta^f - \chi_3^m\varsigma_3(\vartheta^f - 1)}{\vartheta^f - \varsigma_3(\vartheta^f - 1)},
$$

$$
\chi_3^c = \frac{\chi_3^f\vartheta^f - \chi_3^m\varsigma_4(\vartheta^f - 1)}{\vartheta^f - \varsigma_4(\vartheta^f - 1)}.
$$

$$(B.14)$$

B.3 Derivation of Parameters for Hill's Yield Surface

In the following, the parameters of the Tsai-Wu yield condition are related to the parameters of Hill's yield condition. In this context, the expressions related to the Hill's equation have been extracted from BANABIC, 2010.

In case the difference between compressive and tensile yield stress is omitted, the Tsai-Wu yield condition (3.58) for an orthotropic material simplifies to

$$
\begin{aligned}
\Phi = {} & F_{11}\sigma_1^2 + F_{22}\sigma_2^2 + F_{33}\sigma_3^2 \\
& + 2F_{12}\sigma_1\sigma_2 + 2F_{13}\sigma_1\sigma_3 + 2F_{23}\sigma_2\sigma_3 \\
& + F_{44}\sigma_4^2 + F_{55}\sigma_5^2 + F_{66}\sigma_6^2 \qquad\qquad = 1.
\end{aligned}
\tag{B.15}
$$

Such an expression is e.g. the result if a von Mises yield surface is inserted as the matrix yield formulation in the expressions given in section 5.3.1.

In the following, all variables related to the Hill yield condition are labelled by the superscript H. The Hill yield condition given in (3.57) can be expressed as

$$
\begin{aligned}
\Phi^H = {} & \xi_1^H(\sigma_2 - \sigma_3)^2 + \xi_2^H(\sigma_3 - \sigma_1)^2 + \xi_3^H(\sigma_1 - \sigma_2)^2 \\
& + 2\xi_4^H\sigma_4^2 + 2\xi_5^H\sigma_5^2 + 2\xi_6^H\sigma_6^2 \\
= {} & \left(\xi_1^H + \xi_3^H\right)\sigma_2^2 + \left(\xi_1^H + \xi_2^H\right)\sigma_3^2 + \left(\xi_2^H + \xi_3^H\right)\sigma_1^2 \\
& - 2\xi_1^H\sigma_2\sigma_3 - 2\xi_2^H\sigma_1\sigma_3 - 2\xi_3^H\sigma_1\sigma_2 \\
& + 2\xi_4^H\sigma_4^2 + 2\xi_5^H\sigma_5^2 + 2\xi_6^H\sigma_6^2 \qquad\qquad = 1.
\end{aligned}
\tag{B.16}
$$

As opposed to equation (B.15), the interaction terms in expression (B.16) are interrelated. This results in nine independent parameters for the orthotropic Tsai-Wu yield surface in equation (B.15), while merely six parameters are necessary for the Hill definition from expression (B.16).

When the simplified Tsai-Wu definition of the yield surface is given and a material model with a Hill yield surface needs to be parametrised, the parameters of (B.15) have to be transferred to (B.16).

From expression (B.16) it can be determined that the following relations apply for the shear related parameters:

$$
\xi_4^H = \frac{F_{44}}{2} = \frac{1}{2\left(Y_{23}^s\right)^2}, \quad \xi_5^H = \frac{F_{55}}{2} = \frac{1}{2\left(Y_{13}^s\right)^2}, \quad \xi_6^H = \frac{F_{66}}{2} = \frac{1}{2\left(Y_{12}^s\right)^2}.
\tag{B.17}
$$

The variables Y_{23}^s, Y_{13}^s and Y_{12}^s represent the shear related yield stresses.

For uniaxial stress states the yield strength values Y_1^t, Y_2^t and Y_3^t can be extracted from equation (B.15) by application of expression (5.54). These yield strengths can then be used to derive the remaining Hill's yield surface parameters as:

$$\xi_1^H = \frac{1}{2}\left(\frac{1}{(Y_2^t)^2} + \frac{1}{(Y_3^t)^2} - \frac{1}{(Y_1^t)^2}\right),$$

$$\xi_2^H = \frac{1}{2}\left(\frac{1}{(Y_3^t)^2} + \frac{1}{(Y_1^t)^2} - \frac{1}{(Y_2^t)^2}\right), \qquad (B.18)$$

$$\xi_3^H = \frac{1}{2}\left(\frac{1}{(Y_1^t)^2} + \frac{1}{(Y_2^t)^2} - \frac{1}{(Y_3^t)^2}\right).$$

The Hill yield condition of (B.16) can be modified to be fulfilled when a reference yield stress σ_Y is met

$$
\begin{aligned}
\sigma_Y = &\left[(\sigma_Y)^2\left(\xi_1^H + \xi_3^H\right)\sigma_2^2 + (\sigma_Y)^2\left(\xi_1^H + \xi_2^H\right)\sigma_3^2 + (\sigma_Y)^2\left(\xi_2^H + \xi_3^H\right)\sigma_1^2 \right.\\
&- 2(\sigma_Y)^2\,\xi_1^H\sigma_2\sigma_3 - 2(\sigma_Y)^2\,\xi_2^H\sigma_1\sigma_3 - 2(\sigma_Y)^2\,\xi_3^H\sigma_1\sigma_2 \\
&\left. + 2(\sigma_Y)^2\,\xi_4^H\sigma_4^2 + 2(\sigma_Y)^2\,\xi_5^H\sigma_5^2 + 2(\sigma_Y)^2\,\xi_6^H\sigma_6^2\right]^{1/2} \\
= &\left[\left(\bar{\xi}_1^H + \bar{\xi}_3^H\right)\sigma_2^2 + \left(\bar{\xi}_1^H + \bar{\xi}_2^H\right)\sigma_3^2 + \left(\bar{\xi}_2^H + \bar{\xi}_3^H\right)\sigma_1^2 \right.\\
&- 2\bar{\xi}_1^H\sigma_2\sigma_3 - 2\bar{\xi}_2^H\sigma_1\sigma_3 - 2\bar{\xi}_3^H\sigma_1\sigma_2 \\
&\left. + 2\bar{\xi}_4^H\sigma_4^2 + 2\bar{\xi}_5^H\sigma_5^2 + 2\bar{\xi}_6^H\sigma_6^2\right]^{1/2}
\end{aligned}
\qquad (B.19)
$$

The stress computed by the right hand side of the yield condition in (B.19) can be interpreted as effective stress.

Applying the results of (B.18) and the so called R values, the parameters of (B.19) can be expressed as

$$\bar{\xi}_1^H = \frac{1}{2}\left(\frac{(\sigma_Y)^2}{(Y_2^t)^2} + \frac{(\sigma_Y)^2}{(Y_3^t)^2} - \frac{(\sigma_Y)^2}{(Y_1^t)^2}\right) = \frac{1}{2}\left(\frac{1}{R_{22}^2} + \frac{1}{R_{33}^2} - \frac{1}{R_{11}^2}\right),$$

$$\bar{\xi}_2^H = \frac{1}{2}\left(\frac{(\sigma_Y)^2}{(Y_3^t)^2} + \frac{(\sigma_Y)^2}{(Y_1^t)^2} - \frac{(\sigma_Y)^2}{(Y_2^t)^2}\right) = \frac{1}{2}\left(\frac{1}{R_{33}^2} + \frac{1}{R_{11}^2} - \frac{1}{R_{22}^2}\right), \qquad (B.20)$$

$$\bar{\xi}_3^H = \frac{1}{2}\left(\frac{(\sigma_Y)^2}{(Y_1^t)^2} + \frac{(\sigma_Y)^2}{(Y_2^t)^2} - \frac{(\sigma_Y)^2}{(Y_3^t)^2}\right) = \frac{1}{2}\left(\frac{1}{R_{11}^2} + \frac{1}{R_{22}^2} - \frac{1}{R_{33}^2}\right)$$

and

$$\overline{\xi}_4^H = \frac{(\sigma_Y)^2}{2(Y_{23}^s)^2} = \frac{3}{2R_{23}^2},$$

$$\overline{\xi}_5^H = \frac{(\sigma_Y)^2}{2(Y_{13}^s)^2} = \frac{3}{2R_{13}^2}, \tag{B.21}$$

$$\overline{\xi}_6^H = \frac{(\sigma_Y)^2}{2(Y_{12}^s)^2} = \frac{3}{2R_{12}^2}.$$

Hence, in case the Hill condition is implemented in a material model in terms of the R values, these can be derived from expressions (B.20) and (B.21) by

$$R_{11} = \frac{Y_1^t}{\sigma_Y}, \ R_{22} = \frac{Y_2^t}{\sigma_Y}, \ R_{33} = \frac{Y_3^t}{\sigma_Y},$$

$$R_{23} = \frac{Y_{23}^s}{\sigma_Y/\sqrt{3}}, \ R_{13} = \frac{Y_{13}^s}{\sigma_Y/\sqrt{3}} \ \text{and} \ R_{12} = \frac{Y_{12}^s}{\sigma_Y/\sqrt{3}}. \tag{B.22}$$

Usually the reference stress σ_Y is set to Y_1^t which represents the yield stress at $0°$ uniaxial tensile loading in the principal axis. Hence, in such a loading case, the yield condition will be met when Y_1^t is reached. Thus, the value R_{11} resolves to 1.

For the case of plane stress, the Hill equation (B.16) results to

$$\Phi^{H,\text{pl.stress}} = \left(\xi_1^H + \xi_3^H\right)\sigma_2^2 + \left(\xi_2^H + \xi_3^H\right)\sigma_1^2$$
$$- 2\xi_3^H\sigma_1\sigma_2 + 2\xi_6^H\sigma_6^2 = 1. \tag{B.23}$$

Alternatively, the coefficients of the Hill yield condition for the plane stress case are often given in form of the r values, respectively Lankford's coefficients as

$$r_0 = \frac{\xi_3^H}{\xi_2^H},$$

$$r_{90} = \frac{\xi_3^H}{\xi_1^H} \ \text{and} \tag{B.24}$$

$$r_{45} = \frac{\xi_6^H}{\xi_1^H + \xi_2^H} - \frac{1}{2}.$$

The values r_0 and r_{90} can thus be expressed by yield stresses. If the yield stress at $45°$ tensile loading $Y^t_{45°}$ is known, r_{45} can be calculated from r_0 and r_{90}, cf. BANABIC, 2010. This is expressed by

$$r_0 = \frac{\frac{1}{\left(Y^t_1\right)^2} + \frac{1}{\left(Y^t_2\right)^2} - \frac{1}{\left(Y^t_3\right)^2}}{\frac{1}{\left(Y^t_3\right)^2} + \frac{1}{\left(Y^t_1\right)^2} - \frac{1}{\left(Y^t_2\right)^2}},$$

$$r_{90} = \frac{\frac{1}{\left(Y^t_1\right)^2} + \frac{1}{\left(Y^t_2\right)^2} - \frac{1}{\left(Y^t_3\right)^2}}{\frac{1}{\left(Y^t_2\right)^2} + \frac{1}{\left(Y^t_3\right)^2} - \frac{1}{\left(Y^t_1\right)^2}} \quad \text{and}$$

(B.25)

$$r_{45} = \left(\frac{2}{\left(\frac{Y^t_{45°}}{Y^t_1}\right)^2} - \frac{1}{2} + \frac{r_0}{1+r_0} - \frac{r_0(1+r_{90})}{2r_{90}(1+r_0)}\right)\frac{r_{90}(1+r_0)}{r_{90}+r_0} - \frac{1}{2}.$$

In literature related to the modelling of SFRTP materials, the r values are often referred to as fitting parameters that need to be adjusted so that the experimentally acquired curves are met by the Hill model. This is problematic since, as presented in (B.25), these values are not independent of each other. However, the relations in (B.25) show that if experimental tensile curves at $0°$, $45°$ and $90°$ are available the only actually unknown value is Y^t_3. Therefore, finding the appropriate r values can be simplified by merely interpreting Y^t_3 as a fitting parameter in a reverse engineering application. However, by applying the constitutive model proposed in this work the necessary yield strengths can be directly determined from expression (5.54).

Appendix C

Additional Results for Carbon Fibre Reinforced Thermoplastics

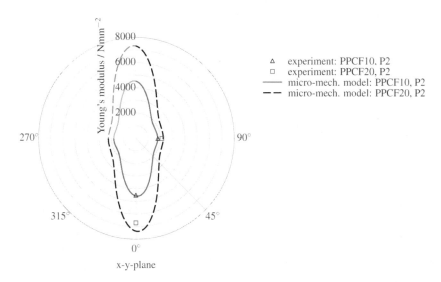

Figure C.1 Polar plot of micro-mechanical results for the Young's modulus of PPCF10 and PPCF20.

Printed in the United States
By Bookmasters